SEVEN GLORIOUS DAYS

OTHER BOOKS BY KARL W. GIBERSON

Worlds Apart: The Unholy War Between Science & Religion

Species of Origins: America's Search for a Creation Story
(with Donald Yerxa)

Oracles of Science: Celebrity Scientists versus God and Religion
(with Mariano Artigas)

Saving Darwin: How to Be a Christian and Believe in Evolution

The Language of Science & Faith:
Straight Answers to Genuine Questions
(with Francis Collins)

Quantum Leap: How John Polkinghorne
Found God in Science and Religion
(with Dean Nelson)

The Anointed: Evangelical Truth in a Secular Age
(with Randall Stephens)

The Wonder of the Universe: Hints of God
in Our Fine-Tuned World

Saving Adam (forthcoming)

SEVEN GLORIOUS DAYS

A Scientist Retells the Genesis Creation Story

Karl W. Giberson

PARACLETE PRESS
BREWSTER, MASSACHUSETTS

2012 First Printing

Seven Glorious Days: A Scientist Retells the Genesis Creation Story

ISBN 978-1-55725-928-8

Library of Congress Cataloging-in-Publication Data

Giberson, Karl.

Seven glorious days : a scientist retells the Genesis creation story / Karl W. Giberson.

p. cm.

Includes bibliographical references (p.).

ISBN 978-1-55725-928-8 (trade pbk.)

1. Creationism. 2. Creation—Biblical teaching. 3. Bible. O.T. Genesis I—Criticism, interpretation, etc. 4. Religion and science. I. Title.

BS651.G485 2012

231.7'652—dc23

2012016029

10 9 8 7 6 5 4 3 2 1

Published by Paraclete Press
Brewster, Massachusetts
www.paracletepress.com
Printed in the United States of America

To the students

from that amazing

Contemporary Questions class[1]

who still remind me

that teaching young people

is a great way to live in

the fantasy that you are not

getting older.

CONTENTS

SEVEN GLORIOUS DAYS

The Seven Days, or Epochs, of Creation

DAY 1

In the beginning God created all that is. The Logos of creation, out of which the heavens and the earth and all things within them burst forth, was the pattern of God's purpose from which everything would emerge and toward which everything would evolve.

And there was evening and morning, beginning and ending, of the first epoch of creation.

And God saw that it was Good.

DAY 2

Then God said, "Let matter emerge, with precisely defined properties that will empower the development of everything else in the universe, laying a secure foundation for changes that will eventually lead to living creatures, following the patterns laid down by the Logos."

And there was evening and morning, beginning and ending, of the second epoch of creation.

And God saw that it was Good.

DAY 3

Then God said, "Let the matter be gathered into stars that they may shine forth, creating life-giving chemicals, providing light to decorate the night and energy to sustain future plant life. And let these stars create gravitational centers about which planets can safely revolve, following the patterns laid down by the Logos."

And there was evening and morning, beginning and ending, of the third epoch of creation.

And God saw that it was Good.

DAY 4

Then God said, "Let planets of every imaginable size and composition emerge, traveling in precise and predictable orbits about their suns. And let some of these planets be located in temperate zones around their suns, covered with water over which my spirit will hover and call forth life, following the patterns laid down by the Logos."

And there was evening and morning, beginning and ending, of the fourth epoch of creation.

And God saw that it was Good.

DAY 5

Then God said, "Let the waters bring forth living creatures in abundance. Let all manner of living things fill the seas and move onto the land and into the air. Let life find, create, and occupy every imaginable niche, maximizing the variety and diversity of possibilities. And let all creatures be empowered to reproduce and respond creatively to changes in their environments, following the patterns laid down by the Logos."

And there was evening and morning, beginning and ending, of the fifth epoch of creation.

And God saw that it was Good.

DAY 6

Then God said, "Let life grow steadily in its capacity to understand and experience the world at the deepest level and to discover the nature and origins of its own existence. And let life develop its own creative capacities so that the world may be filled with great novelty and rich experiences for all its creatures, following the patterns laid down by the Logos."

And there was evening and morning, beginning and ending, of the sixth epoch of creation.

And God saw that it was Good.

DAY 7

Then God said, "Let the members of the species *Homo sapiens* grow to understand the meaning, power, and significance of love; let them understand the importance of right and wrong. And let them burn with a deep spiritual hunger to know the God that created them and the world they inhabit. Let them begin to understand the mystery of the Logos that lies at the heart of their existence."

And there was evening and morning, beginning and ending, of the seventh epoch of creation.

And, having entered into fellowship with creatures in the universe, God blessed the creation and rested, satisfied that the Logos, still present and active in the cosmos, was accomplishing its task.

And God saw that it was Good.

INTRODUCTION

The Creation Story for the Twenty-First Century

"In the beginning God created the heavens and the earth."

I grew up in a religious tradition that taught me to read the Bible literally and accept its statements—on all subjects—at face value. Not surprisingly, I became an enthusiastic young-earth creationist, convinced that the story of origins in Genesis was an accurate scientific account of how things came to be. I headed off to college in 1975 hoping to become a champion of creationism and a fearless crusader against scientific theories like the big bang and evolution—man-made theories that foolishly presumed to challenge the story that God had provided in the Bible.

While studying science at Eastern Nazarene College near Boston I became convinced that the scientific explanations for our origins were true. The biblical account, read literally, simply could not be reconciled with the facts that science was discovering. It had to be reinterpreted. Like many Christians who discover this, I found a way to move past my biblical literalism. I accepted what my "liberal" professors were saying—namely, that the Genesis story was best understood as a theological statement about the relationship between God and the Creation, not a modern scientific account. This timeless theological truth, unfortunately, was presented in the context of an ancient worldview that contained ideas about nature—like the sky being solid or the moon having its own light—that we have long since abandoned.

The Bible is an ancient text with wisdom that still speaks to us today. But we must exercise care in not trying to find modern science in its various references to the natural world. The Gutenberg Bible, pictured here, was the first book printed by a machine.

Preserving the ancient science of the Genesis creation story in the face of what science has discovered makes it look tarnished, like it is showing its age. The story's once-vibrant colors can seem faded and scratched by the clinical, mathematical, and coldly rational tools of modern science. However, the secular scientific story we live with today has lost its power to move us because it seems opaque and impersonal, with no natural place for us. Purpose has been banished from the science of today just as surely as Adam and Eve were banished from Eden long ago. The unchanging character of the biblical text—once celebrated as timeless, ancient wisdom into the ways things are—appears defeated by the impermeable objectivity of a science that looks to the future for wisdom, not the past. But humans need more than a mechanical story about how the gears and pulleys of the cosmos pulled life up out of the primordial ooze. We need a robust, full-orbed creation story, with a place for us.

Since my college days, which were followed by a stint earning a PhD, I have spent over a quarter century teaching science to Christian college students. I have come to appreciate just how tragic it is that so many young people believe they have to accept the ancient worldview in Genesis as an accurate description of the world of today—as if the science of the Bible was millennia ahead of its own authors. So many times I have wondered what the Genesis story would look like if we could update it with the science of today—replacing the seven days of creation with billions of years and recasting the whole story in the context of modern science.

If we could do this, it would be easier for us to believe that God is the Creator. The more we tie God's creative work to the ancient worldview in Genesis, the less likely people are to take the idea of a creator God seriously. Unfortunately, there is no legitimate way to "update" the Genesis creation stories, so we are stuck with this problem and an endless controversy about our origins. Unless . . .

I offer this book as a literary exercise in what the Genesis story might look like if we *could* update it with the wisdom and latest understanding gained from modern science.

I have presumed to rewrite the text of the first chapter in the Bible, which I understand is expressly forbidden somewhere in the book of Revelation. In so doing I am bringing modern science to bear on the story of creation. I have also reshaped the scientific story of our origins as if it was the story of how God created the world, and not merely an account of natural history. This means, of course, that a purposeful thread runs through natural history from the moment of big bang into the present and beyond.

Retelling the scientific story was an exercise in selection. I wanted this to be a short book so that readers would not find themselves lost on long side roads about the formation of stars or how the earth ended up with liquid water. I had to pick those parts of the scientific story that fit together most naturally with *our* story—how our remarkable species came to be. I thus discuss at length how atoms originated—since we are made of atoms—but I say nothing about the origin of galaxies or black holes, which play a more limited role in the origin of our species. I also don't take up controversial topics like the existence of life on other planets, or the possibility of other universes. If there are other universes, or creatures in other parts of this universe, then they have their own creation stories. I am interested only in *our* story.

The biblical creation story says nothing about *how* God actually creates. "Let there be . . . " is presented as an authoritative divine command that brings things into existence. The science of the last couple centuries has established that nature is filled with *processes* that do creative work. In the pages that follow, I have thus located God's creativity *within* the natural order, collaborating with those laws, rather than working *apart* from them. If parts of my discussion seem to omit reference to God it is only because I intend the reader to understand that God is working *through* all of the processes. I think it is important that we do not treat nature as if God does some things supernaturally while allowing the rest of nature to proceed on its own, without God's involvement.

To resonate with the biblical account, I have divided creation into seven "epochs," each with a major theme and each representing an important stage in the history of the universe. Because there is a well-developed—but profoundly misguided—interpretive tradition that treats the "Days of Creation" as cosmic and geological epochs, I hasten to emphasize that this is not the basis for my treatment. There is no scientific basis whatsoever for a seven-fold division. What I present here, however, is at least consistent with how the scientific story appears in terms of the *sequence* of events. As the epochs unfold I highlight the relevance of the trajectory as creation moves toward human life and its remarkable capacities for love. Not surprisingly, perhaps, I treat the emergence of love in its many manifestations as the central purpose of creation.

The story told in these pages is inspired by the Judeo-Christian creation story, but I do not intend by this emphasis to belittle or demean other creation stories. This book is about the Judeo-Christian story in the light of contemporary science and does not venture outside that paradigm. Other creation stories deserve their own treatments from writers familiar with those traditions.

And finally, I have titled the book *Seven Glorious Days* to emphasize my conviction that we live in a world that is truly glorious and, while imperfect and not without its dark corners, is something we should celebrate, cherish, and look after.

The story is summarized here, alternating between the new "biblical" description and the scientific content of each of the seven epochs of creation. And then I fill in the details in the chapters that follow.

FIRST DAY OF CREATION

In the beginning God created all that is. The Logos of creation, out of which the heavens and the earth and all things within them burst forth, was the pattern of God's purpose from which everything would emerge and toward which everything would evolve.

The universe begins with a mystery called, for lack of a better term—and there once was a contest to find a better term—the *big bang*. The moment of the big bang remains, after decades of exploration and theorizing, tantalizingly beyond the grasp of our science. We cannot observe it directly, and even our theories can only take us *close* to that moment but stop short, like Moses looking into the Promised Land but not being allowed to enter. Even our simulations of the early universe in high energy laboratory settings can't get us back to that point. What we can do, though, is see what emerges from the big bang, and our simulations and theories start working amazingly well just a fraction of a second after that mysterious and inaccessible moment of creation.

And there was evening and morning, beginning and ending, of the first epoch of creation.
And God saw that it was Good.

SECOND DAY OF CREATION

Then God said, "Let matter emerge, with precisely defined properties that will empower the development of everything else in the universe, laying a secure foundation for changes that will eventually lead to living creatures, following the patterns laid down by the Logos."

In the second epoch of creation atoms and molecules appear—the building blocks of all that is to come, from stars and galaxies to planets and people. Atoms are endlessly recycled, for they can be used again and again with no deterioration of any sort. A hydrogen atom freely floating in space ten billion years ago can journey to earth on the asteroid that drove the dinosaurs to extinction seventy million years ago. The same atom can be a part of Moses' physical makeup. It can be on the vinegar offered to Jesus on the cross and in rain that fell during the Crusades. Today it can be in a cloud overhead, and tomorrow it can be in a child's squirt gun.

And there was evening and morning, beginning and ending, of the second epoch of creation.
And God saw that it was Good.

THIRD DAY OF CREATION

Then God said, "Let the matter be gathered into stars that they may shine forth, creating life-giving chemicals, providing light to decorate the night and energy to sustain future plant life. And let these stars create gravitational centers about which planets can safely revolve, following the patterns laid down by the Logos."

At a critical point in the early history of the universe, after innumerable hydrogen atoms have been gathered by gravity into huge spherical clouds, the hydrogen ignites. Across the universe these great clouds of hydrogen cross an amazing threshold and become stars, bursting forth in the blackness of space like fireworks on Independence Day; gravity has made the clouds so dense that their atoms are literally crushed together until they fuse with each other, like separate drops of mercury forming into globs. Here we discover one of the many astonishing processes in nature—the mechanism by which simple atoms combine into larger and more complicated ones that make life possible.

The process by which stars shine builds the periodic table as hydrogen atoms combine with each other, and then these combinations join with other combinations and so on. The process can seem chaotic and random up close, but from far away we see that this is how the raw materials of life are created, and the periodic table of the elements is filled in.

Toward the end of their long lives, some of the largest stars can become overwhelmed by their own gravity and undergo catastrophic inward collapses. This process is so violent that the stars actually "bounce" and explode with the force of a billion atomic bombs. Such explosions, called supernovas, populate vast regions of space with the diverse elements created inside the star by billions of years of fusion; the explosions are strangely orderly and eerily silent since empty space carries no sound. Gravity steps up to gather the stellar material back into big clouds again. A large cloud at the center of the explosion can become another, second generation, star. Our sun is one such star.

And there was evening and morning, beginning and ending, of the third epoch of creation.
And God saw that it was Good.

FOURTH DAY OF CREATION

Then God said, "Let planets of every imaginable size and composition emerge, traveling in precise and predictable orbits about their suns. And let some of these planets be located in temperate zones around their suns, covered with water over which my spirit will hover and call forth life, following the patterns laid down by the Logos."

Complex arrangements of atoms called *molecules* have been built from simple raw materials; a universe that was once nothing but vast swaths of hydrogen gas now has solar systems where chemically rich planets orbit about stars with remarkably stable outputs of light. Planets orbiting at just the right distance from their "suns" have a temperature where water is liquid.

Liquid water is rare beyond belief in the universe as a whole—so rare that one would hardly even list it as a component on an inventory. But water is plentiful on our planet and has become a rich and central part of our lives and our cultures: children love to play in it, adults enjoy the beauty of lakes and ponds, and it plays a sacramental role in our rituals, such as baptism.

And there was evening and morning, beginning and ending, of the fourth epoch of creation.
And God saw that it was Good.

FIFTH DAY OF CREATION

Then God said, "Let the waters bring forth living creatures in abundance. Let all manner of living things fill the seas and move onto the land and into the air. Let life find, create, and occupy every imaginable niche, maximizing the variety and diversity of possibilities. And let all creatures be empowered to reproduce and respond creatively to changes in their environments, following the patterns laid down by the Logos."

Liquid water and complex chemical raw materials brought the universe to life. In some extraordinary sense, we can now speak in meaningful terms about the universe having *information*—tiny blueprints directing the formation of ever more interesting and varied forms of simple life.

Subtle interactions between these life-forms as they compete for resources make them increasingly more robust, as the fitter ones reproduce more effectively. The copying process steadily and mysteriously pushes life-forms to greater and greater complexity. The information molecule driving all this would one day be identified as DNA. This remarkable chemical structure possesses amazing abilities to reliably make copies of itself and to explore small variations in the details of those reproductions. These explorations would allow the molecule to locate small improvements to its basic structure and then reproduce that new variation with greater efficiency until it would come to dominate.

A major change occurred when single-celled forms of life began to cooperate and form multi-celled organisms. This cooperation empowered new developments that led to increases in sophistication. Specialized functions appeared, enabling organisms to collect visual information, to hear sounds, to enjoy constant body temperatures, to have solid skeletal structures that would provide protection when they were on the outside and great mechanical dexterity when they were on the inside.

We celebrate the kinds of things that make our sensory experience of the world wondrous, that enable the universe to experience itself. These achievements are major, not trivial, developments.

And there was evening and morning, beginning and ending, of the fifth epoch of creation.
And God saw that it was Good.

SIXTH DAY OF CREATION

Then God said, "Let life grow steadily in its capacity to understand and experience the world at the deepest level and to discover the nature and origins of its own existence. And let life develop its own creative capacities so that the world may be filled with great novelty and rich experiences for all its creatures, following the patterns laid down by the Logos."

As complexity increased, the need to process information increased also, and a remarkable central processing unit of enormous power and sophistication emerged. These *brains*, as they would one day be called, endowed their possessors with a growing capacity to function in the world and to understand the world.

Mysteriously, these brains that evolved in response to challenges having to do with mundane priorities like survival and reproduction acquired capacities to think about complex subjects. The capacity to do mathematics emerged and with it came increasingly deep insights into the patterns and underlying order of creation. Art and music appeared. The universe learned to sing and praise and celebrate. God's creativity was enlarged as creatures began their own process of creation.

And there was evening and morning, beginning and ending, of the sixth epoch of creation.
And God saw that it was Good.

SEVENTH DAY OF CREATION

Then God said, "Let the members of the species Homo sapiens *grow to understand the meaning, power, and significance of love; let them understand the importance of right and wrong. And let them burn with a deep spiritual hunger to know the God that created them and the world they inhabit. Let them begin to understand the mystery of the Logos that lies at the heart of their existence."*

Eventually the most advanced of the life-forms on the planet, human beings, became deeply religious. Throughout the history of our species, belief in God or gods has been close to universal. Abstractions such as right and wrong inspired reflections on how we should live. Questions about the meaning of life would occupy our deepest thinkers. And trying to understand where everything came from would become a critically important question pursued by every human culture. The religious impulse developed into one of the deepest aspects of our complicated understanding of ourselves.

And there was evening and morning, beginning and ending, of the seventh epoch of creation.

And, having entered into fellowship with creatures in the universe, God blessed the creation and rested, satisfied that the Logos, still present and active in the cosmos, was accomplishing its task.

And God saw that it was Good.

DAY 1

In the Beginning

In the beginning God created all that is. The Logos of creation, out of which the heavens and the earth and all things within them burst forth, was the pattern of God's purpose from which everything would emerge and toward which everything would evolve.

Once upon a time . . .

How many grand stories begin with this familiar phrase? Once upon a time there was a little girl named Goldilocks. . . . Once upon a time there was a gentleman who married, for his second wife, the proudest and most haughty woman that was ever seen. . . . Once upon a time there was a little girl who lived in a village near the forest. . . .

Fairy tales begin so innocently. There is nothing wondrous about a little girl who lives in a village near the forest. I grew up in such a village and live in another one now. My grown daughters were once "little girls who lived in a village near a forest." But we have learned from that most sacred of places—our mother's knee—that journeys that begin, "Once upon a time," will surely take us someplace interesting. And the fairy tales do not disappoint. The ordinary little girl who lives in the ordinary village on the edge of an ordinary forest soon encounters witches, wolves, giants, talking bears, and pipers with magical flutes.

Our universe began "once upon a time." Like the fairy tales, there was a defined starting point about which the story is silent. We are not told what went before or how the little girl came to live in the village near the forest. We understand that there is some kind of a "world" in place, with rules and patterns that must be followed: people live in villages; witches are bad; Good triumphs over Evil; and little girls have adventures that usually end well. But within the world defined by these simple rules we know that a captivating story is about to unfold and, when it ends, we will have learned something interesting.

Our universe began with a mystery. There is no science of such "beginnings." Nobody has a laboratory where tiny universes are coaxed into existence to see how the process works. Our largest telescopes cannot peer into some far-off corner and watch universes popping into existence, like soap bubbles blown into the wind by laughing children. Our science is all about the behavior of universes once they get going. We have nothing but the vaguest of hints as to how they originate in the first place. It is a deep mystery and one that engages many of our greatest scientific minds.

This mystery has a name—the big bang. It's an uninspiring label, if you think about it. It originated as an expression of disrespect but, like the eggs that hooligans throw at houses on Halloween, it stuck. A corny sitcom about physics graduate students has taken *The Big Bang Theory* as its name and, I must confess, the label seems more appropriate for lowbrow comedy fare than for the grand narrative of our universe.

The moment of creation remains out of reach, even of our imaginations. We have no theories, computer simulations, or models of what it means to say that a universe is born.

The term *big bang* appeared in 1949 at a time when few scientists were taking the idea seriously. A flamboyant and antireligious astronomer named Fred Hoyle introduced it on a BBC radio show. Hoyle was suspicious of speculation that the universe had originated "once upon a time" in some kind of magical eruption beyond the reach of science. Such a beginning implied, well, a *beginning* to the universe. Hoyle thought this looked way too much like the first verse of the Bible. "In the beginning God created the heavens and the earth."

Hoyle had reason to be suspicious of the seemingly magical story. The idea he was dismantling for his BBC listeners was the brainchild of a Roman Catholic priest named Georges Lemaître. Hoyle thought the church had made up all kinds of magical stories over the centuries and now here they were again making up another one.

Lemaître, however, was far more than a priest. He was one of Belgium's leading scientists. He taught at the Catholic University of Leuven and was one of Europe's leading experts on the fledgling science of the cosmos. And he was quite insistent that his ideas about the universe were the result of his *science*, not his *religion*. He even scolded the pope for suggesting that astronomers were starting to glimpse the moment of creation.

The skeptical Hoyle was not so sure. He suspected Lemaître had constructed a rather clever Trojan horse and was slowly wheeling theology through the gates of science with a hypothesis he dismissed as the big bang. Even Lemaître's religious colleagues had problems with where he was going. The great Quaker astronomer and collaborator of Einstein, Sir Arthur Eddington, was concerned, even though he had made some important contributions to the very science Lemaître was exploring. "Philosophically," Eddington wrote in 1931, "the notion of a beginning to the present order of nature is repugnant to me."[2] Two years later, as the "notion of a beginning" continued to take shape on the horizon, he reiterated his preference in distinctly nonscientific terms: "the most satisfactory theory," he wrote in his popular work *The Expanding Universe*, "would be one which made the beginning not too *unaesthetically abrupt*."[3]

Hoyle's dismissive label for the "unaesthetically abrupt" beginning to our universe is still with us. It has become such a part of our vocabulary that we no longer notice that science has appropriated a phrase from the gutter. But compare *big bang* to its counterparts in other branches of physics—*classical dynamics*, *quantum mechanics*, *nuclear physics*. Doesn't it look oddly out of place on such a list? Cosmologist and science writer Brian Swimme renamed the big bang the "Primordial Flaring Forth" in the subtitle to one of his books.[4] The new name never caught on.

There once was a contest to find a better term. A band of astronomy buffs, led by acclaimed science writer Timothy Ferris, held a competition to come up with a more flattering name for that mysterious "moment of creation." The existing label was actually a misnomer, they complained. The event that birthed the universe "wasn't big and didn't go bang."[5] The results of the contest were announced at a 1994 meeting of the American Astronomical Society in Washington, D.C. Carl Sagan was one of the judges.

More than ten thousand entries, from more than forty nations, had been submitted for review. Proposals included "The Trip from Zip," "Blast from the Past," "Spark in the Dark," "First Fireball," "The Primordial Poof," "Cosmic Kaboom," and even "Buddha's Burp." "The Creation" was a popular option, not surprisingly, and perhaps confirming Hoyle's suspicion that something theological was hiding in the theory. "The winner is . . . no one," announced Ferris at the close of the contest. "None can surpass the term 'Big Bang.'" And that is why the central theory of cosmology is not known as "Buddha's Burp."

So, while the name of the theory is inelegant—and likely to stay that way—the theory itself is breathtaking. The actual *moment* of the big bang remains elusive, and there is little hope that this will change. We cannot observe this moment directly, and we are not likely to develop a good enough "guess" to have any confidence we have it right. But what we *can* do is see the results after the fact and, like the clever detectives on television, we can assemble enough clues to work our way backward to see what happened in the grandest of historical exercises.

Science has been quite successful at traveling backward through time. Over the past century the grand cosmic narrative has been written, one chapter at a time, starting with the one in which we are living today. We now understand that our solar system, including the sun, originated about five billion years ago from a large cloud of atoms and molecules that were gathered into huge balls by gravity. This cloud was created when a star in this part of the Milky Way exploded. This original star originated some ten billion years ago, in much the same way that our sun originated five billion years ago.

Gravity creates stars by gently tugging on hydrogen atoms until they gather into huge balls. These balls got larger and larger until their gravitational pulls became so strong they started to crush the atoms out of which they were made. This crushing process, amazingly, causes the stars to ignite through a process known as *fusion*—gigantic ongoing nuclear explosions that can last for billions of years. Fusion compresses the simple hydrogen atoms so powerfully that they combine into larger atoms, like little gobs of mercury joining together into a bigger gob. Atoms like helium, carbon, nitrogen, and iron originated through this fusion process. Larger atoms—carbon, nitrogen, oxygen, iron—make life possible. Life cannot be constructed in any form from hydrogen. We now understand the remarkable truth that all the atoms in our bodies today were fused in this ancient furnace.

This is an amazing perspective. Look at your hands. Imagine you can see enough detail to distinguish the individual atoms. Now look at the sun or another star if it is night. *The atoms in your body were once inside a star*. The billion-year journey they took from that star to your hands is amazing. It is the most provocative of the many plotlines that run through the grand narrative of creation.

This is not a tale that science tells very well. It ends up boring the astronomy students who have to learn it for the upcoming exam. Poets like Joni Mitchell do it much better. I used to play her beautiful song "Woodstock"—a big hit for Crosby, Stills, Nash and Young—for my students, calling attention to the lines where she says, "We are stardust and" and "We are billion-years-old carbon."

The hydrogen atoms that gravity long ago gathered into stars—destined to become the stardust of which we are made—were created in an earlier chapter in the history of the universe. Electrically charged protons and electrons, whizzing about in a cosmic electrical storm, were attracted to each other and gradually "found" each other, like a mother and child separated in a busy parking lot. Things calmed down and simple hydrogen atoms formed from the unions. These simple atoms, made of a single negatively charged electron orbiting about a single positively charged proton, were electrically neutral, their equal and opposite charges canceling each other out. This allowed gravity, a force much weaker than electricity, to begin its slow work of gathering these particles into gigantic balls to make stars and eventually stardust. Before the charged particles combined into neutral atoms, the electrical force dominated gravity.

The protons at the center of the hydrogen atoms were themselves composed of even smaller particles called *quarks,* the most recent additions that physicists have added to the growing family of particles in the universe. Three such quarks have to combine in just the right way to make a proton.

This "quark era" was yet an earlier chapter in the grand narrative of the universe. It was also the first chapter in the history of the universe with physical matter playing a lead role. But this is not the first chapter of the story.

The quark era embodies what can only be called the physical incarnation of the rational foundations—or the Logos—of the universe. It is here, in the Logos of the universe, that our ability to keep traveling back in time comes to a close. We are like climbers who have been climbing a hill that grows ever steeper and disappears in front of us into the clouds above; less hardy climbers drop out along the way; many stay behind, satisfied with the terrain at lower altitudes, or skeptical that the top of the mountain is attainable. The heroic climbers who have forged upward in faith finally end up on a ledge looking up at a sheer cliff that disappears into clouds far above their heads. From this vantage point we wonder if we will ever find a way to get onto the next plateau.

The agnostic astronomer Robert Jastrow captured this sentiment in a passage straight out of Fred Hoyle's nightmares:

> For the scientist who has lived by his faith in the power of reason, the story ends like a bad dream. He has scaled the mountains of ignorance; he is about to conquer the highest peak; as he pulls himself over the final rock, he is greeted by a band of theologians who have been sitting there for centuries.[6]

This powerful image captures the conviction that gradually settles into the souls of cosmologists as they work their way back to the beginning of the universe. What little they can glimpse of the beginning looks nothing like the rest of the story. The physical reality of the world, with its comforting and secure tangibility, disappears into clouds of pure mathematics. The messiness of a world cluttered with molecules, planets, stars, and galaxies slowly evaporates and is replaced by what looks—if we could only see it—like a beautiful mathematical symmetry.

The moment of the big bang—that elusive prelude to "once upon a time"—defines the laws and patterns that shape the development of the universe. It's like a parent presenting children with a grand set of Legos, knowing that the very structure of the Legos will both enable and constrain the children as they construct grand and imaginative castles. Out of the big bang come the Legos of the universe. This rational undercarriage, hidden within, behind, and underneath the physical reality that we see today, guides the universe through the many chapters of its history: quarks are turned into protons, protons into atoms, atoms into stars, stars into solar systems with planets, chemicals into life, and life into us. And then we, in appreciation, invent science and uncover the story of how it all happened.

The deepest and most fundamental laws of physics, with their various and wide-ranging properties, constrain, enable, and specify the things that can happen. Nothing occurs outside these laws, from the crashing of an asteroid into the earth to the recollection of a childhood memory; from the explosion of a nuclear bomb to the coloring of the sky at sunset; from the radioactivity at a nuclear waste site to the precipitation of a summer rain. Nothing happens in the universe without the "official permission" of the laws of physics.

Remarkably only *four* different kinds of things can happen in nature. Every event, no matter how simple or complex, results from some combination of these four different types of events. Oftentimes the events are so complex it is hard to locate the "physics" inside them, but it is there, nonetheless, as surely as your beating heart hides inside your chest.

Our mythology speaks of Four Horsemen of the Apocalypse that will bring about the end of the Creation. We can recast this metaphor more optimistically to be the Four Horsemen of the Creation that come riding out of the mystery of our origins to empower the processes that generate the structures of the universe.

The most familiar, and by far the weakest, of the Four Horsemen is *gravity*. Locally, this ubiquitous if feeble fellow keeps us in our chairs and prevents our atmosphere from drifting off into space. It keeps the earth orbiting about the sun and the tides moving in and out. Over the history of the universe gravity collected material into stars and planets, and created all the large-scale structures in the universe today—planetary systems, galaxies, galactic clusters, and more.

Also familiar is the horseman of *electromagnetism.* This powerful engine of creation runs our life processes. The chemical reactions in our bodies that convert food into energy to power motion and even thinking are guided by electromagnetic interactions between atoms and molecules. The process, when viewed from afar, is astonishing. Light from the distant sun falls on green grass, where it is absorbed and utilized by the chlorophyll molecule to create stored energy. Cows eat this grass, turning it into meat. Then we eat the meat, and the stored energy is released to power our activities. Indirectly, but surely, our energy comes from sunlight. Reflect on that the next time you have a meal.

Sunlight is produced by a third horseman known as the *strong nuclear force.* This is the most powerful of the forces, as its name suggests, and plays two major roles in making life possible. The first is the most obvious: the production of sunlight. The strong force converts matter into light as a by-product of fusion reactions that take place inside stars, where lighter atoms, such as hydrogen and helium, are fused into heavier atoms, such as carbon and oxygen. The force is called "strong" because it can fuse two protons together, despite the powerful repulsion they feel for each other. Protons are all positively charged and, like powerful magnets that push apart when you try to force the same poles to touch, they resist being close to each other. When the strong force makes larger atoms out of smaller ones, some of the matter is converted into energy, just like in a nuclear reactor. The energy is released as sunlight, which heads off into space. A tiny fraction of this sunlight, of course, hits the earth.

The second way the third horseman contributes to life is by producing the complex materials on which life is based. Hydrogen and helium are too simple—chemically boring in ordinary language—to make anything interesting. Imagine trying to build something out of marbles. For structures as complex as life, we need atoms like carbon that can connect to other atoms. But no carbon was created in the big bang. In fact, no carbon was present anywhere in the universe for millions of years. The carbon on which life is based was fused in stars under the relentless effort of the strong force, steadily building heavier elements from lighter ones. What an amazing synergy that life-sustaining sunlight is produced in the same process that makes the chemicals of life.

The final of the four horsemen is the least appreciated—the so-called *weak nuclear force*. This force performs an odd function—it turns neutrons into protons. And this process releases some dangerous radiation, which is why real estate is such a bargain next door to nuclear waste sites. But the weak force also helps the sun create its radiation and so is critically important to life.

The zigzag path by which simple hydrogen atoms combine into heavier elements and release energy contains many complicated steps. And there are lots of rules that must be followed along the way. One especially critical step occurring early in the process requires that the weak force transform a proton into a neutron. This step absolutely must occur if the universe is not to remain sterile and lifeless.

After two protons fuse in the first step along this path, the new created duo will not fuse smoothly with another two-proton combination. Protons don't want to be together in the first place, so getting four of them—two pairs—to fuse in the face of strong electrical repulsion is just too hard. But if one of the protons is turned into a neutron—creating a variation of hydrogen called *deuterium*—then this combination will fuse more readily. By turning one of the protons in the duo into a neutron, the charge on the combination is cut in half, thus reducing the repulsion it experiences from other positively charged combinations. This, of course, makes it easier for them to "run into each other," and fuse. But even this picture is a bit too simple.

One of the many rules that ride into the universe with the four horsemen is the conservation of electrical charge. The total charge on—in?—the entire universe appears to be *zero*, and there is a strict rule specifying that it must stay that way. For every positive charge in existence there is a negative one to balance it. If you want to make a new positive charge, you have to find a way to simultaneously make a negative one to cancel it out. Atoms, of course, have no charge, since the number of protons in their nucleus is equal to the number of electrons orbiting about the nucleus. The sum total of all the positive and negative charges in an ordinary atom is, thus, zero. So when the weak force converts a proton into a neutron, it cannot just strip away the positive charge and toss it aside, as though it were a banana peel. The positive charge has to "go" somewhere. What happens is amazing.

The positive charge disappears from the proton and reappears on a tiny particle called a *positron*, which is created by this transformation to preserve the charge. A positron is the antimatter version of an electron—in reality as well as science fiction. The most likely future for a positron is to randomly encounter a stray electron and explode in mutual annihilation, releasing a burst of energy. In the science-fiction television series *Star Trek*, this is the process that powers the warp engines on the *Enterprise.* In the real world, this energetic explosion helps keep the sun shining, and shining, and shining, for billions and billions of years.

The Four Horsemen of the Creation work both singly and as a team. Gravity's construction of stars and planets is the best example of a solitary project. In contrast, fusion in stars is a joint effort. The four horsemen work together to pull that carriage up the long hill to the bottom of the periodic table of the elements.

Remarkably, everything described above is mainstream science, accepted by all cosmologists. None of it is speculative or controversial. Introductory textbooks present these ideas to freshman astronomy students in college. They are the foundations of the science of cosmology.

The interactions of the Four Horsemen of the Creation are stunning. A universe that began as little more than a sea of blinding energy has unfolded with the narrative grace of a literary classic. And now, billions of years later, as we look at cosmic history, we can feel in our bones that something extraordinary has occurred, that cosmic history has not been a fourteen-billion-years-long tale told by an idiot, full of sound and fury but signifying nothing.

Freeman Dyson, one of the great scientists of the twentieth century, had an unusually poetic soul and a scary grasp of advanced mathematics. In his autobiography, *Disturbing the Universe*, he reflected on what the universe looked like to him when he stepped back and pondered it all at once:

> The more I examine the universe and the details of its architecture, the more evidence I find that the universe in some sense must have known we were coming.[7]

This was not a scientific inference, and some of Dyson's colleagues wondered what he was thinking to make such an odd and seemingly meaningless claim. Dyson endorses no particular religious viewpoint—he calls himself a "practicing but not believing" Christian—so he was not advancing a faith perspective. He was simply affirming, as a scientist who understands that the world is more than the sum total of its scientific explanation, that he feels at home in the universe. He was acknowledging the Logos of creation.

As we look back to the beginning of the universe through the lenses of our largest telescopes and more profound theories, we are, to be sure, "seeing through a glass darkly." And the glass gets darker the further back we go. In fits and starts we unravel the mystery of our origins, figuring out what happened "once upon a time." The understanding we seek is an apparition in the fog, a shiny object under the water, a song we can just barely hear, an unexpected footprint on an isolated shore.

A footprint is a provocative image. When Robinson Crusoe, after years of isolation, found Friday's footprint, it altered his lonely little world. There could be but one explanation for that footprint, and he was energized to understand it. Like Crusoe, we too are gazing at a footprint and wondering. In fact, we are gazing at a long series of footprints that start on our beaches and invite us to trace them back to the beginning of time.

"We have found a strange footprint on the shores of the unknown," wrote Eddington, when the fledgling science of cosmology was in its youth and the term *big bang* had not yet been coined. "We have devised profound theories, one after another, to account for its origin. At last, we have succeeded in reconstructing the creature that made the footprint. And lo! It is our own."[8]

And there was evening and morning, beginning and ending, of the first epoch of creation.

And God saw that it was Good.

DAY 2

A Universe of Horseshoe Nails

Then God said, "Let matter emerge, with precisely defined properties that will empower the development of everything else in the universe, laying a secure foundation for changes that will eventually lead to living creatures, following the patterns laid down by the Logos."

A popular poem, its origins lost in European folk traditions, speaks of the way small events can have great consequences:

For want of a nail the shoe was lost.
For want of a shoe the horse was lost.
For want of a horse the rider was lost.
For want of a rider the battle was lost.
For want of a battle the kingdom was lost.
And all for the want of a horseshoe nail.

We use "luck" to explain events that have no explanations. "How lucky you are to win the lottery!" we say enviously to our friend. "I had such bad luck today," I complained recently to my wife. "The police pulled me over when I was going the same speed as everyone else." My students say, "Good luck on your exam," to each other as they arrive at my class on exam day.

The chance character of events we call "lucky" makes them mysterious, which is why dice games are so much fun. Something beyond our control determines the numbers on our dice, and we watch with great anticipation as the dice come to rest on the table in front of us. In a more serious vein, we read in the Bible of people casting lots as a way to discern God's will. The "lots" were stones that would be rolled and then examined to see which one seemed to glow with God's favor. People believed God could speak through the casting of lots, since influencing the outcome was beyond human ability. Proverbs 16:33 explains the reasoning: "The lot is cast into the lap, but the decision is wholly from the Lord."

Lots were cast throughout the Old and New Testaments. One of the more dramatic occasions was the decision of the disciples to replace Judas. In Acts 1:24–26, we read: "And they prayed and said, 'Lord, who knowest the hearts of all men, show which one of these two thou hast chosen to take the place in this ministry and apostleship from which Judas turned aside, to go to his own place.' And they cast lots for them; and the lot fell on Matthias; and he was enrolled with the eleven apostles."

Luck did not fare well in the centuries after it helped select a replacement for Judas. You might even say it was unlucky. By the eighteenth century science had genuine luck on the run as physicists like Isaac Newton proclaimed with great authority that all objects move in accordance with strict laws of motion. Luck, by the emerging lights of the new science, had absolutely no role in determining the result of a tossed coin. The result of the toss is completely determined by factors like air pressure, the angular and linear momentum imparted by your thumb when the coin is tossed, wind, how far it falls, and other related but entirely mechanical factors. A well-designed machine, steadily flipping identical coins in a vacuum, could flip an unbroken sequence of heads for as long as you wanted—no luck involved, and rather boring to watch.

The uncertainty—the mysterious "luck"—surrounding coins and dice comes from our ignorance of the things that influence them, not from some mysterious "luck" that intrudes into our world from another beyond this one. Mechanical objects behave in mechanical ways. This perspective was all well and good and didn't bother anyone until they realized that, as near as science could tell—and everyone believed that science could tell quite a bit—we are all mechanical objects that must follow the same laws that apply to coins being flipped.

This curious and, some would say, *dismal* picture was expressed eloquently by the French physicist Pierre-Simon Laplace. As the eighteenth century came to a close he wrote the follow obituary for "luck," effectively banishing it from the universe.

> We may regard the present state of the universe as the effect of its past and the cause of its future. An intellect which at a certain moment would know all forces that set nature in motion, and all positions of all items of which nature is composed, if this intellect were also vast enough to submit these data to analysis, it would embrace in a single formula the movements of the greatest bodies of the universe and those of the tiniest atom; for such an intellect nothing would be uncertain and the future just like the past would be present before its eyes.[9]

The universe, as understood by Laplace, was fully determined from the beginning. Only because we cannot identify all the relevant details is the universe at all interesting. If we could, we would know that we are but billiard balls on a cosmic table, moving because we were struck in the past by another ball, relevant because we strike some other ball so the motion continues. The past, present, and future merge into one gigantic pattern with all the details specified.

The nineteenth century lived with Laplace's gloomy determinism, but just as that century was about to close, physicists began to spot what looked like luck, sprouting like weeds in the vast neatly manicured lawns of determinism. In what I think is the single deepest insight that science has acquired to date, genuine luck—physicists call it *indeterminism*—was discovered hiding behind the tiniest particles. The theory that brought this shy behavior into the light of day was called *quantum mechanics.*

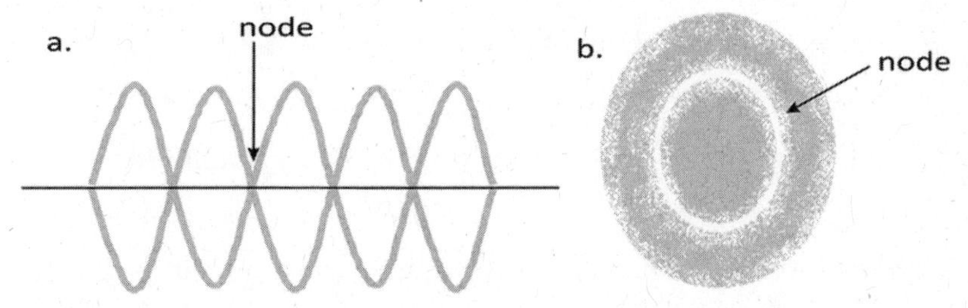

Quantum mechanics reveals a world where particles like electrons behave as if they have a "free" choice about where they are going to be. Electrons have a probability distribution that specifies the likelihood they will be found in a particular place. In figure "a" above, the electron will most likely be found where the wave is at its highest level, but there is no way of knowing which of the peaks the electron will be at. It will almost never be found at the nodes of the wave. Figure "b" shows the probability distribution in an atom. The electron can be found in the shaded parts but never in the node. Figure "b" illustrates the paradox where an electron can move from one region to another without ever being observed in between the two regions.

Like most scientific revolutions that topple long-standing and seemingly secure dominoes, quantum theory was heroically opposed by leading physicists. The evidence suggested that particles, including the familiar electrons, had something that looked a lot like free will. They acted nothing like the billiard balls that Laplace envisioned when he banished luck from the universe. How could this be?

The "free will" of the electrons is impossible to understand in any conventional sense, but we can observe it. Electrons approaching a grid with holes in it, we have discovered, can seemingly choose which hole they pass through. Identical electrons make distinctly nonidentical choices, like the same coin being tossed in the same way but giving different results. Or people ordering different food in a restaurant.

These discoveries were all very peculiar; books written about this era have titles such as *The Strange Story of the Quantum*, *Thirty Years that Shook Physics*, or *Alice in Quantumland*. Some physicists, most notably Albert Einstein, completely rejected the new ideas. "God does not play dice with the universe," the great physicist said, in an oft-quoted objection to the new theory. In print and at conferences all over Europe, Einstein argued about quantum mechanics with Niels Bohr, the great Danish physicist who presided over the dismantling of Laplace's universe. "Stop telling God what to do," Bohr scolded Einstein.

The Bohr-Einstein debate is perhaps the greatest intellectual battle ever fought in science. The question was deep, and the sparring titans were the world's two leading scientists at the time. Bohr—tall, powerful, charismatic, and endlessly talkative—insisted that physicists had no choice but to embrace the strange world of the quantum where electrons appeared free to do things of their own choosing. Einstein—eccentric, reflective, and unfathomably brilliant—acknowledged Bohr's arguments but could not bring himself to accept that actual genuine "luck" was anything other than an illusion or an indication that we don't understand things very well. What, he wondered, could this mean? On what basis would two identical electrons in identical situations "decide" to do two different things? Why would one go through the hole on the left and the other go through the hole on the right? Surely the world only *appeared* to be like this; this couldn't be the way it actually *was.*

Einstein never accepted the quantum picture of the world, even as Bohr brought the entire physics community around to the new view. Sitting quietly, lost in thought in his own untethered canoe, Einstein floated away by himself and became an almost solitary voice objecting to the strange new world opened up by the quantum revolution.

By the time the dust settled, an Austrian physicist named Erwin Schrödinger had developed a mathematical equation to describe what these "free" electrons were doing. His equation showed how electrons were guided in their course by a "probability wave" that kept them from going too far afield, while letting them be completely free on a small scale—like a dog running about on a leash, or a child in a playpen.

In the decades since Bohr pushed Einstein out to sea, quantum theory has come to be viewed as the deepest thing that physicists know about the universe. As reality is peeled back and we look at smaller and smaller pieces, we eventually discover a world completely dominated by the quantum picture of reality. Even the so-called vacuum of space has a quantum character, with particles—seemingly freely—popping in and out of existence. The quantum vacuum unpredictably creates temporary particles that live for the briefest of moments and then disappear back into the no-man's-land from which they came, as if they were briefly sampling what the universe was like and then deciding that they didn't like it. Such peculiar quantum behaviors are found everywhere in the realm of the very small.

The quantum laws that rule the world are a deep and central part of the Logos of creation. Everything we know suggests that they were here first, "in the beginning." The particles that appeared later had no choice but to obey these laws, just as a person leaping from an airplane must do as gravity commands. The quantum laws are even deeper than the Four Horsemen of the Creation that we looked at in chapter 1, for they are the laws that the Horsemen must obey. Gravity and electromagnetism can indeed gather particles together, but the quantum laws specify how they are to accomplish that task.

When the universe crossed that mysterious threshold and came into being, it brought no particles with it. The popular picture of an exploding ball dispersing the building blocks of stars and planets into empty space isn't even close to what actually occurred. The universe came across that bridge from nowhere to somewhere with little more than a collection of mathematical rules—a cosmic charter—for what kinds of things can happen: energy and electrical charge must be conserved in all cases; quantum rules apply to everything; space can stretch and expand; nothing gets to go backward in time, or faster than the speed of light.[10]

Picture the early universe as a game with unknown rules, which is a good metaphor for how scientists do their work. What kinds of things happen? What patterns are regular? What kinds of things never happen? If several things can happen, which is more common? If the universe were a grand game of chess that we were watching for the first time we could quickly discover the rules: bishops move on their diagonals; the queen can move as far as she wants in any direction; losing your king means the game is over; putting your opponent's king in danger obligates you to say, "Check."

As we watch the universe begin its long journey from "once upon a time," we learn the rules of that game, which eventually becomes our game. Remarkably there are only two kinds of physical particles that appear on the grand board of our game: quarks and leptons. Everything in the universe, from the grass under our feet to the stars over our heads, is made from various types of quarks and leptons. We are made of quarks and leptons. The familiar protons and neutrons are composed of quarks; the electron is the best-known example of a lepton.

All natural phenomena, from the sublime to the mundane, result from two kinds of particles interacting via four kinds of interaction. The grand chessboard of nature allows just two different types of players. But these two players can do many different things: They can disappear in a puff of energy. They can appear, seemingly out of nowhere. They can laze about in small groups in empty space, doing nothing. They can combine into atoms and molecules. Quarks and leptons, however, can't do everything, and as we watch them, we discover the rules they must obey. No interaction—no matter how messy or explosive—creates or destroys electrical charge. Electrical charge is the same yesterday, today, and forever. Its total value in the universe is probably zero—every positively charged particle has a negatively charged counterpart. Mass, however, is not conserved, and collections of particles are constantly engaging in behaviors that alter the total mass of the group. The relevant rule when it comes to mass is $E = mc^2$. This tidy little formula tells mass and energy how to relate to each other. If mass is reduced or destroyed, energy must appear to balance that loss. If particles with mass are to appear out of nowhere, as they often do, energy must be used up. In fact, the best way to understand mass is as bundled energy, with this formula giving the amount of energy that would be bundled in any given glob of mass. And, since the formula tells us to multiply mass by the square of the speed of light—a gigantic number—there is an unbelievable amount of energy tied up in matter.

This famous formula—one of the deepest and most wide-ranging rules in the universe—was discovered by Einstein at the dawn of the twentieth century as the "Thirty Years that Shook Physics" was getting underway. The formula was a central feature of his revolutionary theory of relativity that overturned some of Isaac Newton's ideas that had long been viewed as unassailable.

Three decades later, a brilliant but unsung Hungarian physicist named Leo Szilard came to see Einstein at his home on Long Island. By then Einstein was a famous celebrity, acclaimed as our species' greatest genius, and hounded endlessly by reporters for his opinion on everything from Zionism to women's fashions. Szilard convinced Einstein that his famous formula could be used to build an atomic bomb and that Hitler had possibly started a project to do just that. Einstein put his celebrity to great use and penned a now-famous letter to President Roosevelt that led to the Manhattan Project and the development of the bombs that were dropped on Japan in World War II. Hiroshima and Nagasaki were destroyed by the conversion of mass into energy. Much of the mass that obliterated the two cities originated billions of years ago in the reverse process, as energy was converted into mass. The destructive power unleashed on Japan in 1945 had been bundled into mass by nature fourteen billion years ago at the dawn of the universe, when the laws of quantum mechanics summoned quarks and leptons into existence.

Quarks obey an interesting collection of rules. Even though they have fractional electrical charges—plus or minus 1/3 and 2/3—they must partner with each other to create combination particles with electrical charges of 0 or 1—like a social engagement where dates are required. The familiar proton at the center of every hydrogen atom consists of three particles—two so-called up quarks and one down quark. The up quarks have charges of 2/3 and the down quark has a charge of –1/3. The combination, therefore, has a charge of 1, following the rule: the total charge must be 0 or 1. A neutron consists of two down quarks and one up quark for a total charge of 0, again following the rule.

Quarks owe their odd labels to the linguistic eccentricities of Murray Gell-Mann, who won the Nobel prize in 1969 for figuring out the rules of the quark game. Gell-Mann was a brilliant mathematician and an enthusiastic student of language. He was determined to avoid what he thought were the linguistic blunders of his predecessors, who named things inappropriately based on incomplete knowledge. Atoms, for example, got their name from the Greek *atomos*, which means "indivisible." The label was originally applied to particles thought to be the smallest possible (and thus indivisible); but we have since learned better. Splitting the atom is routine now, so the label is officially a misnomer—atoms are not indivisible. *Planet* is another misnomer, coming from a term meaning "wandering star." Gell-Mann the linguist was unsettled by the many misnomers that Gell-Mann the physicist had to put up with in his day-to-day work. So in crafting the language of these peculiar new particles, he resolved to label things so that no advances in the future would result in his picture ever being hung on the linguistics wall of shame.

These odd new particles—which some leading physicists were calling "Aces"—combined most commonly in trios to form protons and neutrons, although there were also some combinations of just two quarks. This inspired Gell-Mann to borrow a line from James Joyce's inscrutable classic *Finnegan's Wake*. That layered, opaque, and stress-inducing tale contains a scene where some drunks are singing, "Three quarks for Muster Mark!"—a phrase with no clear meaning. "That's it," said Gell-Mann. "Three quarks make a neutron or a proton." Since *quark* had no meaning—as far as anyone could tell—using that label for these newly discovered particles posed no risk from future discoveries that might change the way quarks were viewed. "The whole thing is just a gag," Gell-Mann reflected in a 1983 interview. "It's a reaction against pretentious scientific language."[11]

Gell-Mann's linguistic romp continued as he labeled the rest of the quarks—there were six in all—and determined the rules they had to follow. The six quarks he labeled *up*, *down*, *top*, *bottom*, *strange*, and *charm*—none of which had any meaning whatsoever in the context of quarks. There is no "upness" associated with the up quark that might be threatened by future scientific advances. Upness, downness, topness, bottomness, strangeness, and charmness were completely immune to correction based on future scientific advances because they referred, in a *literal* sense, to nothing that had anything to do with quarks. Gell-Mann's idiosyncratic labeling has led to some even stranger phrases in physics, such as a law known as the "conservation of strangeness," which invites jokes about the people who use such phrases.

If Gell-Mann the linguist was a playful fellow, Gell-Mann the mathematical physicist was anything but. The quark theory that he developed, with others, in the 1960s is based on the belief that reality rests on the most elegant of mathematical foundations. The entire theory of quarks was inspired by the way that the hundreds of elementary particles in nature organized themselves. Particles that seemed otherwise unrelated could be brought into a harmonious pattern if they were composed of a few simpler particles. Just as we understand that every atom is made from electrons, protons, and neutrons, so all the elementary particles—the less familiar mesons, hadrons, and others—looked like they could be constructed from a short and simple roster of six quarks.

We now have a reasonably clear picture of the long journey that matter has taken from the dawn of time into the present. The universe crosses that threshold into being as little more than a set of mathematical rules specifying what can and cannot happen. The deep quantum character of this embryonic reality compels it to explore possibilities—to try its luck, so to speak, and see what happens. In less time than the blink of an eye, things start to happen. Quarks and leptons appear, their existence paid for by losses in the bath of energy that brings them into existence. The quarks, with their charges of 2/3 and –1/3, combine under the influence of the strong nuclear force. Soon they are all comfortably gathered into protons and neutrons, passively obedient to the requirement that fractionally charged particles cannot exist on their own. Everything in the universe now has a charge of 0 or 1.

At the same time, negatively charged electrons are being born, always careful to keep the total charge of the universe at 0. They buzz about, with the protons and neutrons, in a chaotic but steadily cooling mix as the universe expands. Mathematical rules that Gell-Mann and others would discover fourteen billion years later have laid the first foundations for the material structure of the universe. The weak and strong nuclear forces have guided the unruly collection of particles into their now familiar forms: protons, neutrons, and electrons.

A second stage of construction begins. The same force that binds the quarks together inside the protons and neutrons, reaches outside those particles and grabs quarks in other particles. The force is short range—like Velcro—but adequate to grab on to the quarks inside a particle if it gets close enough. The most common example of this is the occasional proton that gets stuck to a neutron. And, in rare circumstances, two protons might get stuck together, but as their electrical charges create a mutual repulsion, they usually avoid each other.

The strong force is most important at the beginning of the universe, when the protons and neutrons are being created. As the universe cools and the buzz of particles subsides, the torch of creation is passed to the electrical force, which pulls the electrons toward the protons. As the temperature declines across a certain key threshold, the electrons suddenly drop into orbits around the protons and the universe is full of hydrogen atoms. The laws of quantum mechanics specify exactly where the available orbits are located. The electron cannot orbit anywhere it pleases. But there are options, and the electron has freedom to choose one orbit over another.

All the particles in the universe are now electrically neutral; most of them are hydrogen atoms. And, since electrically neutral particles are not attracted to each other, the powerful electrical force is no longer relevant. Once again the torch of creation is passed on to yet another horseman, this time to the gravitational force. Although gravity is by far the weakest force, and tiny hydrogen atoms don't tug on each other with much enthusiasm, there are millions of years for gravity to work its magic. And it does.

Hydrogen atoms are pulled together by gravity ever so slowly. They begin to cluster, and the clusters grow faster because they have more mass. And eventually the clusters become stars, where the next stage of creation takes place. Hydrogen atoms are fused into heavier and heavier atoms, such as helium, carbon, oxygen, and the like. The periodic table of the elements is created inside stars. And, when some of those stars explode as supernovas, these elements are made available for gravity to turn into planets with rich chemical possibilities. Our earth is one such planet.

All this history is a necessary prerequisite to the world we enjoy today. The atoms in our bodies were once inside a star. Many of the hydrogen atoms in our bodies—and we are mostly composed of water, which has two hydrogen atoms in every molecule—were formed in the early stages of the universe. Those hydrogen atoms have taken an amazing journey to get to where they are right now. Consider the possible route that a hydrogen atom may have taken to get into your body. Thirteen billion years ago it was floating freely in space, having recently been formed by an electron dropping into an orbit around a proton. Ten billion years ago it was inside a star, buzzing about frantically in the midst of the ongoing reactions that make the stars shine. Six billion years ago that star exploded and the hydrogen atom rocketed off into space, only to be captured by gravity and become a part of a planet that was forming 93 million miles from a new star, which was being created from the exploded debris of an old one. Three billion years ago that atom was part of a waterfall cascading down on our planet when there was no life more complex than a single cell. Seventy million years ago the atom was a part of the chemistry of a dinosaur on the verge of extinction. Two thousand years ago that same atom could have been present in a bead of sweat on Jesus's forehead as he ran about playing with his friends under the hot sun while his mother, Mary, made dinner. The atom may then have spent a millennium in the oceans of the world and traveled from one continent to another in various streams. And now that atom is in your body.

We cannot know, of course, what our atoms have been up to. They don't carry passports or leave trails in the sands of time. But all of them are billions of years old and have done many interesting things before they came to be a part of our makeup. Atoms and molecules are nature's great recycling project, used again and again over the course of billions of years but never worn out. Unlike, say, paper, which deteriorates with each successive recycling, the laws of physics keep atoms in their original state for billions of years. Atoms never get old.

The most awe-inspiring aspect of this long, strange trip is the constant presence of mathematical laws, guiding and controlling every aspect. When we examine the world at the "top level," so to speak, the mathematics is invisible. The world outside our window can seem disorganized and chaotic. Children, sunsets, and forests, for example, may all be breathtakingly beautiful, but their beauty does not derive very directly from the mathematical laws that contribute so much to their nature. And there are many processes in nature that on the surface are certainly not beautiful—decaying meat covered with flies, forest fires, or the digestive process. But the hidden mathematical laws that guide such processes are themselves quite beautiful.

The great beauty of mathematics, while apparent to those that study it, is all but invisible to those who do not. In fact, many of us even have negative experiences with mathematics that make it hard to even imagine what mathematical beauty might look like.

Mathematics has much in common with music, and we can use this comparison to create an analogy for what we mean when we speak of mathematical beauty, hidden beneath the surface layers of the world.[12]

Imagine that a friend is taking you on a stroll down a long, seemingly endless, incredibly noisy hallway. As you enter the hallway the noise is deafening—a combination of explosions, metal crumpling, loud music from incompatible genres, babies crying, talk show hosts yelling, and so on. As you wander down the hallway, your friend explains that his company makes filters that eliminate noises, as long as they know how those noises are produced.

He demonstrates the technology for you. As you cross a line marked 10 he turns on a filter that eliminates the sound of explosions. At 9 the crumpling metal racket disappears. By 5 there is just loud talking, babies crying, and discordant music. At 1 there is nothing but beautiful classical music and a loud talk radio host yelling something about his taxes being too high. At 0 the talk radio host is gone. You have come to the end of the hallway and are standing on a balcony on the opposite side of the building looking out into a dark abyss. Beautiful music is coming out of the darkness.

"Impressed?" says your guide, to which you answer, "Of course."

"We don't know anything about this music," says your friend. "It seems to be a sort of beautiful bedrock sound that is just there."

"But aren't you puzzled by this beautiful music?" you ask. "Surely you must have some explanation for it."

"Nope," he says. "I used to wonder about the music but, as you can see"—he gestures into the abyss—"it is just there." You look at him in puzzlement, wondering how something so grand can be taken for granted.

This example makes use of the way we all understand and even identify with music. This familiarity helps us appreciate the inadequacy of simply accepting the existence of beautiful music with no explanation for its origin. Unfortunately, few of us have any idea what it might mean to describe mathematics as beautiful and even less an idea about the mystery raised by its existence.

In this analogy we simply replace "noise and sounds" in the story above with "nature" and replace the beautiful music coming from the abyss with the mathematical equations that physicists have discovered at the foundations of reality. On the surface, nature is, to be sure, noisy in the sense of being cluttered, busy, and seemingly without patterns. Even beautiful scenery—picture a mountain lake with snowcapped mountains in the background—rarely seems "organized." But as we apply our scientific knowledge to the cluttered world we experience and drill down to the bedrock of our understanding—eliminate the noise—we find something quite wondrous. At the end of the great hallway that takes us from the social sciences to the natural sciences, through biology and chemistry and ultimately to physics, we find ourselves at last in the presence of a most beautiful and unexplained symphony of mathematics. Across the dark abyss, this mathematics comes clearly into view, out of nowhere, explaining the world around us while remaining unexplained itself. It is part of the Logos of creation.

A desire to understand the world does not compel all of us to ponder the origin of mathematics. Many of us don't even like math, and I have lost count of how many people have rolled their eyes at me when I told them I was a math major in college. But those that understand it best impulsively lean over the railing into the abyss because they know in their bones that there is something out there.

And there was evening and morning, beginning and ending, of the second epoch of creation.

And God saw that it was Good.

DAY 3

A Billion Stars Are Born

Then God said, "Let the matter be gathered into stars that they may shine forth, creating life-giving chemicals, providing light to decorate the night and energy to sustain future plant life. And let these stars create gravitational centers about which planets can safely revolve, following the patterns laid down by the Logos."

In one of the most beloved stories in all of literature we read of "wise men from the east" following a star until it came to rest over a stable in humble Bethlehem. Every holiday season that star the wise men followed sits symbolically atop our Christmas trees, reminding us of that first Christmas.

Stars have presided over many of the great events of the past, and our folklore is filled with stories of their messages and influences. It's an ancient intuition that lives on in the work of those who continue to weave fanciful tales of how the stars overhead shape our mundane stories here on Earth. Somewhere deep in our souls we hold on to the belief that the world is a gloriously united whole and that everything relates to everything else. Millions of us happily hand our money to gurus who claim they can show us what the stars are telling us about our lives. We are far less interested in what those stars overhead actually are. This is why, in the United States, astrologers are ten times more common than astronomers.[13]

For as long as we have records, humans have found patterns and meaning in the stars overhead. In northern climates the appearance of Orion every fall was a message that winter would soon arrive.

Our species has always lived under the stars and we have a long and enduring relationship with the patterns in the night sky. In a time before light pollution and smog obscured the stars, before the distraction of television, before we regulated our years with calendars and our days with clocks, we marveled at the stars overhead. It was impossible not to see the shapes and patterns that God had placed there for our comfort and guidance. This group of stars looks like a Big Dipper; the group over there looks like a Little Dipper. This collection looks like a hunter with a sword hanging from his belt. Let's call him Orion. And the arrival of Orion in the night sky, we soon learned, accompanied the arrival of winter in northern climates as surely as bears coming out of hibernation signified the arrival of spring.

Our ancestors all saw pictures in the night sky. We find catalogs of the stars from people living thousands of years ago in the valley of the Euphrates River, which the Bible says ran out of the Garden of Eden. The remnants of these ancient records suggest that the ancients saw the lion, the bull, and the scorpion in the stars. These early images were supplemented by the Greeks and Romans, who created much of the sky lore we have today and whose labels adorn the planets.

By the fifth century BC, the constellations were coming to life and began to be viewed as divine. The "wandering stars" received their modern names: Mars, Jupiter, Mercury, Venus, Saturn—each one a god with different influences. The second-century astronomer/astrologer Claudius Ptolemy of Alexandria, one of the greatest thinkers of the first millennium, wrote, "It is

Saturn's quality to cool and, moderately, to dry." Mars's nature was "to dry and to burn." Venus "warms moderately because of her nearness to the sun."[14]

Critics of astrology, which now include the entire scientific community and most educated people, were rare in antiquity. The Roman Cicero, in his book *On Divination*, expressed skepticism about astrology on various grounds. He noted, for example, that the "parental seed" had far more influence on people than the particular stars that looked down on their birth. Twins, born at almost the same time and thus under the same constellations, often led very different lives. Why did the stars bless the life of the firstborn so much more than his brother who arrived a few minutes later?

Cicero's friend, the astrologer Figulus, countered that the celestial sphere carrying the constellations around the earth moved so rapidly that the second twin would indeed have a different sky overhead at his birth. Writing around 400 AD, St. Augustine launched a devastating critique based on this claim, noting that we never know the exact moment of anyone's birth. So, if the celestial sphere turns rapidly overhead, we have no idea what star patterns looked down on the birth. Augustine also noted the story of the famous twins born so close together that the second came out holding the foot of the first—Jacob and Esau. They were born as close together as physically possible and yet experienced dramatically different lives.

Despite such high-powered criticisms, astrology remained the imaginative companion of astronomy—a sort of artsy twin—through the next twelve centuries. For every scholar who rejected astrology, ten embraced it. The star-twins parted during the scientific revolution for reasons more practical than anything else.

Astrology was astronomy's cash cow. Charting the heavens was an expensive enterprise needing observatories, equipment, and assistants. This required patronage of some sort. The value of astronomy by itself was limited. Why, for example, would a king need accurate models of the motion of the stars overhead? Why build expensive and elaborate observatories simply to make better measurements of the stars? What need was there for more precise models of the heavens? Astrology provided a reason. If the fortunes of an empire were linked to the stars, then the king of that empire had better know about those stars. Therefore, most astronomers, including such luminaries as Tycho Brahe, Johannes Kepler, and Galileo Galilei, made part of their living casting horoscopes and advising on the general influences of the heavens. It was a natural complement to the astronomy for which they are famous today.

Careful observation of the heavens provided two distinct kinds of messages: most exciting was the information that could be cast into horoscopes that warned of plagues and hinted at great victories in war. More significant, though, were the scientific messages about the nature and origin of the heavenly bodies and even the entire universe. Nowhere was this dual nature of observation more apparent than in the sudden and startling appearance of a brand-new star in 1572.

New stars were not allowed to appear in the skies over sixteenth-century Europe for several reasons. Those in the largely Christian populace that could read knew that the Bible stated clearly that God had made everything in six days and was now resting. The creation was complete, which meant that no new stars—or planets, or species—would be appearing. This biblical perspective complemented a widely embraced two-thousand-year-old astronomical tradition going back to the Greeks. Aristotle, and others, had insisted that there could be no change of any sort in the heavens. A wandering star, for example, couldn't even change its speed or direction by the smallest amount. Apparent changes, like the apparent backward motion of Mars that occurred when the earth passed it in its orbit, were explained using complex geometrical constructions. Elaborate mechanical models involving wheels within wheels were invented to show how apparent changes in the motion of a planet could occur without there being any actual changes.

The astonishing and predictable regularity of the night sky—an authentic and familiar observation—led to a compelling "law" that the heavens were perfect, complete, and unalterable. Christian thinkers like St. Thomas Aquinas and Dante enthusiastically embraced the idea, even though it originated in Greece centuries before Christ. Theologians had even

Christianized the idea of the changeless heavens by suggesting that the earthly realm was troubled and fallen because of sin, but the "curse" of sin did not extend into the heavens. Hence the glorious perfection of the night sky.

This was the worldview of the Danish astronomer Tycho Brahe when he observed a new star in 1572. Previous observations of interesting "new things under heaven"—like comets or shooting stars—had been dismissed as atmospheric phenomena like wind or rain. Novelties were permitted as long as they were in the earthly realm, which reached to the orbit of the moon but no further. Comets, as we now know, are not in the atmosphere, but nobody knew that in the sixteenth century, so they were mistakenly assumed to be as close as clouds and lightning—which certainly enhanced the threat that doomsayers attached to their appearance.

New stars like the one that appeared in 1572 are rare and none had been observed for centuries. But timing is everything, and Brahe was uniquely situated to take note of this new star and interpret its significance.

Brahe was unquestionably the greatest observer of the pre-telescopic era. Generously supported by the king of Denmark, he had amazing observational facilities on the island of Hven, off the coast of Denmark, in a narrow strait heavily trafficked by boats traveling from the Baltic Sea to the Atlantic Ocean. The island is presently home to a few hundred hardy Swedes. A diligent observer of the heavens, Brahe had equipment that could determine whether a celestial object was closer than the moon, and thus in the earthly realm. If it was beyond the moon, it was in the unchanging and perfect heavens.

Brahe's mysterious new star appeared on November 11, 1572, against the backdrop of the constellation of Cassiopeia. Consistent with prevailing wisdom, other observers assumed that the new heavenly light was in the terrestrial region below the moon. From a purely naked eye perspective there was simply no way to tell. Meteor "showers," to take a familiar example, have long been labeled incorrectly as shooting stars. The lights streaking across the night sky do look like stars that have detached from the firmament, but they are really just rocks that have come into the earth's atmosphere and are burning up.

Brahe knew better. The new star was beyond the moon, even though that was not possible. His equipment and observational skills made that clear. The technique he used to draw this conclusion, called *parallax*, exploited the way nearby objects appear to move back and forth against a distant backdrop if observed from different locations. If you look at your finger while alternately closing one eye and then the other, your finger will appear to move back and forth. Likewise, the moon could be observed, when viewed from different locations, to move back and forth against the backdrop of the stars beyond it, which appear fixed.

Brahe coined the term *nova* to describe what seemed utterly impossible to him: a new star. "Amazed, and as if astonished and stupefied, I stood still, gazing," he wrote about the event. "When I had satisfied myself that no star of that kind had ever shone forth before, I was led into such perplexity by the unbelievability of the thing that I began to doubt the faith of my own eyes."[15]

A new star was a theological puzzle. If God created the heavens "in the beginning," what was this new star? Perhaps it was some kind of message. Speculation began to mount about its astrological significance and the message it contained. In his 1573 treatise on the new star, Brahe the astrologer spoke gravely about impending political upheaval, the like of which had not been seen since the Roman Empire. Rulers born under the sign of Taurus, especially those in Russia, Livonia, Finland, Sweden, and southern Norway, should beware. In between the lines, Brahe the astronomer dropped hints that he knew more than he was divulging—a common strategy for astrologers of the day (and any day, for that matter). After all, they needed more money to keep doing their valuable work.[16]

Not surprisingly, the new star was not a portent of things to come for Taurus Swedes. But Brahe's observation played an important role in undermining the Aristotelian astronomical tradition, a necessary prerequisite for the development of the Copernican system that would succeed it. Far more important, however, is the role that Brahe's new star played in helping us understand the behavior of stars and the amazing role they play in the creation of life in the universe. In ways that Brahe could never have imagined, his new star—and others like it—are critical way stations on the long and winding road from the sterile beginnings of our universe to the gloriously habitable one we find ourselves in today.

The story begins about 300,000 years after the big bang, before there were any stars anywhere in the heavens. The universe was in transition during this epoch of creation, and the processes that produce stars were just beginning to do their work.

We met the strong nuclear force and the electromagnetic force in the previous chapter. These forces are the architects of the atom, determining the character of that important building block of the universe. Under their direction, the charged particles of the early universe gradually combined into atoms. The process took a long time, for the early universe was so energetic that electrons simply could not settle down into stable orbits about protons, just as excited toddlers in preschool don't readily settle down for their afternoon nap.

In the active hurricane of the early universe, the newly minted protons continually tugged on the electrons whizzing by, but to no avail. The light electrons were moving so fast they were long gone before they even knew there were nice neat orbits available for them to drop into.

As the universe expanded, however, the temperature dropped and the electrons slowed to the point where the beckoning protons could nab them as they passed by. Gradually, the electrons were pulled into orbits about the more massive protons, and hydrogen atoms began to appear. Every time an electron joined a proton, two charged particles combined to form an electrically neutral hydrogen atom. The universe slowly lost almost all of its charged particles, a critical first step in inaugurating the next epoch of creation.

The electrically neutral hydrogen atoms were tiny uncharged masses influenced only by gravity, which tugged on them ever so gently.[17] Small clusters of hydrogen formed. Small clusters turned into larger and larger clusters and eventually clouds of hydrogen gas millions of miles across were floating lazily in the pitch-black universe. The clouds grew steadily larger and the hydrogen atoms became more densely packed, which only increased the force of gravity attracting the atoms to each other. Eventually the gravity tugging the atoms pulled them so close together that the hydrogen atoms were compressed into each other. Under massive compression, the nuclei at the center of the atoms got so close together that the strong force came to life and began to pull on the quarks inside the nuclei. The result was fusion—as pairs of atoms transformed into one.

At the critical point where fusion begins, gigantic masses of hydrogen ignite. Across the universe these clouds of hydrogen turn into stars as runaway fusion reactions transform them into nuclear furnaces. The universe, if only there had been observers present, would have seemed like a cosmic fireworks display, as star after star exploded into brilliance against a backdrop of cold, icy darkness.

And here we encounter one of the many astonishing processes in nature—the mechanism by which simple hydrogen atoms are transformed into the more complicated ones that make life possible. As these newly glowing stars shine, they slowly but surely build the periodic table of the elements. The process looks chaotic and random up close—like the maelstrom in the middle of a hurricane—but from far away there is an astonishing degree of order as we watch the raw materials of life slowly appear.

Stars are nuclear explosions that last for billions of years. The process begins with simple hydrogen atoms compressed into larger atoms, like little gobs of mercury joining together into bigger gobs. The larger atoms in the universe today—carbon, nitrogen, oxygen, iron, gold, and silver—were all fused in these ancient furnaces.

As fusion reactions in stars convert hydrogen into helium, the stars grow denser, like a snowball being vigorously packed from a ball of fluffy snow. The increased density compresses the atoms more tightly so that hydrogen fuses even more rapidly and eventually even the helium atoms are compressed to the point of fusion, as they are transformed into carbon, oxygen, and other heavier elements. If the process did not take billions of years, we would call it a "runaway" reaction because the denser the star becomes, the greater the fusion process; and the greater the fusion process, the denser the star becomes.

Stars are incredibly stable, despite being exploding nuclear bombs. Their stability comes from a balance between gravity and the natural expansion of hot gases. Gravity relentlessly compresses everything into a core that gets denser and denser. But this is offset by the tendency of hot gases to expand outward, like the heat and smoke from a fireplace (not to mention the tendency of nuclear bombs to explode!). This balance is not eternal, however, for the fusion process steadily increases the density of the star's core until gravity "wins" the battle and upsets the equilibrium.

Large stars can be overwhelmed by their own gravity and undergo catastrophic inward collapses, like a balloon bursting. These collapses can be so violent that the imploding star can "bounce" and re-explode outward with the force of a billion atomic bombs. When this occurs the star becomes incredibly bright—much brighter than at any time during its normal life. Since most stars are too far away to be visible, this sudden increase in their brightness can make them appear, as if out of nowhere—a "new" star.

The star that puzzled Tycho Brahe in 1572 was just such an exploding star. It had been there all along and, in fact, had actually exploded some 7,500 years earlier. The light from the explosion, which began its journey through space 6,000 years before Christ was born, arrived at the earth on November 11, 1572. The mysterious light was the last gasp of a dying star, not the first appearance of a new one.

Such explosions are known today as *supernovas.* When the stars explode they populate vast regions of space with the elements created inside them, launching a rich chemistry back into the space in which they had formed billions of years earlier; the explosions are strangely orderly and eerily silent since there is no sound in space.

Supernova explosions create the heaviest elements, located on the bottom of the periodic table. Their considerable energies are able to fuse heavy elements like gold and uranium, which are beyond the capacity of the conventional fusion process in stars.

Some five billion years ago a star exploded and spread its stardust throughout a sparsely populated region of a galaxy that would eventually be named the Milky Way. The vast cloud of chemically rich material created by the explosion was slowly—one is tempted to add "and carefully"—gathered by gravity into clumps large and small. As the cloud contracted under the relentless tug of gravity, its temperature rose and it began to spin ever so slowly, and then faster, like a figure skater drawing in her limbs. The rotating amorphous blob of material trembled on the edge of rebirth, as a long-dead fusion reaction began to awaken. The rotating cloud flattened into a disk, like a molten mass of glass spinning at the end of a glassblower's pipe on its way to becoming a plate.

A large mass at the center of the cloud explodes into brilliance. A star is born, or perhaps we should say "reborn." A few billion years later it will be named "the sun." The sudden outward push of energy from this new sun stops the gravitational collapse of the rotating cloud. The clumps of material in orbit about the sun concentrate into a few larger spheres that will one day be known as planets and named after gods. Closest to the sun, Mercury forms from metallic dust too heavy to have been blown away by the ignition of the sun. Venus, Mars, and the Earth form in regions dominated by metals and rocky dust. Earth ends up with significant quantities of water. Beyond Mars, tens of thousands of irregular rocky chunks failed to become a planet. They orbit there to this day, and are known as the asteroid belt. Science-fiction writers weave tales of grand journeys to the asteroids to mine valuable minerals for use back on Earth.

On the other side of the asteroid belt, in what we call the outer solar system, are the large gaseous planets—Jupiter, Saturn, Uranus, and Neptune. They are cold, inhospitable spheres of gas, without solid surfaces.

The embryonic solar system was violent and—if we can use this word in a world without life—dangerous. Massive projectiles—many containing ices of various forms—whizzed about until gravitational tugs from the new planets had them raining down like God's judgment on Sodom and Gomorrah. The earth was bombarded with enough watery projectiles to fill the oceans, a most promising development. The surface of the earth is now constantly refreshed by wind and rain so the scars of this bombardment are long gone, but the process must have been dramatic when it occurred. The weatherless moon has no such process to scour away its history, however, so its craters from long ago are still visible.

Meanwhile, the sun ramped up in the middle of the new solar system, preparing for its role as the overlord of a planetary system and destined one day to be worshiped by millions of creatures. The sun grew steadily hotter and brighter until it reached a steady state about four billion years ago. Eventually we would figure out that the sun converts seven hundred million tons of hydrogen to helium every second—an astonishing rate, but one that can last billions of years. The process is remarkably stable.

The heat from the sun declines steadily as we move outward through the solar system. The surface temperature of Mercury is over 300 degrees Fahrenheit. The side facing the sun is over 800 degrees. The surface temperature of Venus is more than 850 degrees; Mars can be as cold as 200 degrees below zero, and the warmest spot—on the Martian equator—is around the freezing point of water. The planets beyond Mars are all significantly colder.

In between Venus and Mars is a special region where water exists as a liquid. This temperature range is rare in the universe—so rare as to make water in liquid form nonexistent for practical purposes. Smack-dab in the middle of this unusual temperate zone is Earth.

Planet Earth is the ultimate bioengineering success story. The conditions necessary for life are many and precisely specified, but all present on our planet.

The challenges of starting with a universe filled with hydrogen atoms floating about and then arriving at a planet capable of supporting life are overwhelming. Even the most optimistic, clever, and well-educated scientist could *never* have looked at the early universe and predicted that someday it might contain life. This would seem as ridiculous as staring into a box of Legos and wondering if it might somehow give rise to an iPad. Life, as we now understand, is richly complex.

Consider hydrogen—the simplest possible atom. Left to its own devices, hydrogen simply combines with a twin to make a hydrogen molecule. Its social ambitions are easily satisfied by this simple union. A universe filled with hydrogen molecules is no more fun than a playroom with nothing to play with except grains of rice. There just isn't anything very interesting that can be done.

The first step in the direction of life is to create the raw materials, which can be done with fusion. But first you have to make huge stars to create powerful gravitational compressions to get fusion started. Fusion inside a star, however, is a limited process. It can create carbon, nitrogen, oxygen, and other lighter atoms critical to life. But it can't create heavy atoms like potassium, sodium, and iron, not to mention gold and silver. There just isn't enough energy to bring massive atoms with large positive charges into contact with each other. And the useful atoms a star can create are buried inside a raging nuclear inferno that is incredibly hostile to life.

Supernovas, of the sort that mystified Brahe in 1572, solve both problems. The incredible energy of the explosions overcomes the barriers to the fusion of the heavier elements. Materials critical for life are created in the supernova's cauldron of energy as part of the long transformation of the universe from exclusively simple atoms to a rich variety of chemicals.

Imagine a periodic table of the elements chronicling the appearance of various atoms throughout the history of the universe. Imagine that each square begins to glow faintly as atoms belonging to that square appear for the first time. When the universe begins the entire table is black, completely invisible, reflecting the fact that no atoms have yet appeared. Almost immediately the upper left-hand square—hydrogen—begins to glow with blinding brightness. The upper right-hand square—helium—also glows, but dimly. And the third square, just under hydrogen—lithium—has just the faintest of glows, reflecting insignificant traces of that element in the universe.

Nothing changes for hundreds of millions of years, during the long epoch while stars slowly form. But then the stars explode into brilliance all over the universe and the periodic table begins to record the changes. The hydrogen square dims a bit and the helium square comes roaring to life like a newly kindled fire as these first stars fuse helium from hydrogen. Other blocks begin to glow for the first time—beryllium in block 4 and boron in 5. Carbon, in block 6, begins to glow, anomalously brighter than the atoms on either side of it. It turns out that carbon, perhaps the most important atom for life, is not only produced with unusual efficiency, but there is even an odd bottleneck that slows down its fusion into heavier elements.

The squares for nitrogen, oxygen, fluorine, and neon are all glowing brightly now, but carbon stands out. The blocks on the third row glow faintly, including block 14, silicon, just under carbon. Billions of years pass and the blocks on the periodic table begin to glow, some so faintly that they can barely be seen, others very brightly. Some grow steadily as they are fused during the lifetimes of ordinary stars. Others are incremented more unevenly, as byproducts of supernova explosions, or the random and rare collisions of stars.

Atoms inside stars, however, contribute nothing to life, just as food in the grocery store does nothing for hungry people. Remarkably, the same supernova process that fuses some of the heavier elements also spreads its rich pantry of atoms throughout space, where gravity can begin to assemble them into a second-generation star, about which planets can orbit, some of them at exactly the right distance to play host to liquid water.

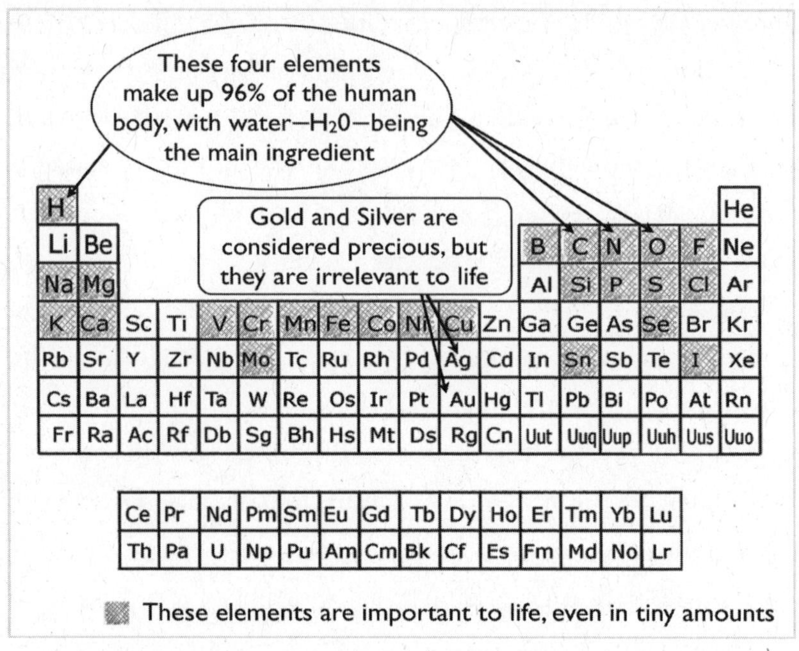

A surprisingly large fraction of the elements on the periodic table are essential to life. Some, like the hydrogen and oxygen in water, are abundant and plentiful in our bodies. Others, like iron and potassium, are present in tiny quantities but are nonetheless critical for health.

The watery planet that would bear the label "Earth" some five billion years later was nothing short of a creative miracle. The ten billion years that it took the universe to produce a habitable planet is about the shortest possible time for that gargantuan task. Skeptics who say humans must be irrelevant because they did not exist for most of the history of the universe don't know what they are talking about. It takes a few billion years to make the first stars and about five billion years for a newly minted first-generation star to fuse itself into a supernova. It then takes a few billion years for the cloud from that supernova to reassemble itself into a second-generation star like our sun, surrounded by rocky planets rich in organic molecules and, in rare cases, liquid water.

If Brahe was appropriately mystified by the supernova of 1572, how much more mystified we should be, as we contemplate the world outside our windows. I write these words from a sunroom on the back of my house that looks into the woods. I am an early riser and often get to see the sun coming up, through the trees in front of me. That sun—a five-billion-year-old second-generation star—is 93 million miles away. It takes eight minutes for the light generated by its fusion reactions to make the long journey from the edge of the sun to my window. Some of that light is absorbed by chlorophyll molecules in the plants outside my window. These molecules contain atoms forged in the nuclear furnace of an ancient star.

Chlorophyll molecules convert sunlight into energy, which can be stored for years in a plant. Some of this energy was in the tomatoes I ate last night. Now that same energy drives my metabolism, keeping me alive, letting me experience this new day, powering my fingers now on my keyboard as I write these words. Outside my window are the flora and fauna of New England—towering maples, azaleas, pachysandra, wild turkeys, deer, countless songbirds at my feeders. Their metabolisms are also powered by the sun. Some of the sunlight warms the ocean after a long New England winter, coaxing summer into existence.

The distant sun literally lights up the scenery outside my window. Everything I see becomes visible only when light strikes it. This same multitasking sun also provides the gravitational force that keeps the earth in its stable orbit, tracing out a mathematically perfect ellipse several billion times in a row.[18]

Planet Earth is our home. And like the homes we played in as children, it is all too easy to take it for granted and find it unremarkable. After all, it is just *there*, under our feet every day. We search diligently for Earth-like planets outside our solar system—places where water can exist in liquid form, making life possible. Such places, we now understand, are rare. When we understand the heroic effort that it takes to create a habitable planet, we can understand why they are not common.

There are many ways to think about our planet. Ever since Copernicus established that the earth was not at the center of the universe, scientists have used the odd term *principle of mediocrity* to describe the earth, suggesting that we must not think of our planet as being special. It is, after all, just a small rocky orb 93 million miles from an ordinary star located in the suburbs of an ordinary galaxy. Carl Sagan used the term *Pale Blue Dot* to describe the earth, noting that from the outer solar system our planet resembles a dust mote, lit up by the rays of the sun. "The earth," he wrote, "is a very small stage in a vast cosmic arena. . . . Our posturings, our imagined self-importance, the delusion that we have some privileged position in the universe, are challenged by this point of pale light. Our planet is a lonely speck in the great enveloping cosmic dark."[19]

Everything that Sagan says is true. The earth is a speck, unimaginably small in a vast universe. But this is only one way to think about our planetary home and perhaps not even a particularly good way. Let us rewrite his sobering words: "The earth is home to the only life we know. On it is a life form that has penetrated the deep mysteries of nature, and laid bare the Logos of creation. To the best of our knowledge, the only thoughts in all this vast universe are located on that pale blue dot. If the universe understands itself in any way, it is only because it brought forth thinking creatures on that dot."

And there was evening and morning, beginning and ending, of the third epoch of creation.

And God saw that it was Good.

DAY 4

Looking for Life in All the Right Places

Then God said, "Let planets of every imaginable size and composition emerge, traveling in precise and predictable orbits about their suns. And let some of these planets be located in temperate zones around their suns, covered with water over which my spirit will hover and call forth life, following the patterns laid down by the Logos."

In January 1610, Galileo pointed his new telescope at Jupiter, the Jovian planet, and noticed some small additional points of light near the giant planet. With no way to know how far away they were he watched them for a week to see if they might be stars, moving in concert with the constellations. Night after night the mysterious lights moved about, but not with the stars. Sometimes there were four lights, sometimes three. Sometimes they were on one side of Jupiter; sometimes on the other. And sometimes they seemed to be hiding behind Jupiter. Always, however, the new points of light were near Jupiter, so Galileo concluded they must be satellites of the planet named after Jove. He recorded the new satellites in his notebook, and started thinking, as he often did, about how to parlay his discovery into political advantage.

The great German astronomer Johannes Kepler, a contemporary of Galileo, was elated by the discovery of satellites around Jupiter. He had recently established, at least to his own satisfaction, that the planets moved in elliptical orbits, rather than circles, as had long been thought. And his mentor, the great Danish astronomer Tycho Brahe, had discovered a new star in 1572, proving that the heavens could occasionally experience some novelty. Kepler had discovered another new star in 1604. Such heavenly novelties raised a lot of questions at the dawn of the seventeenth century, as we saw in the previous chapter.

In reasoning that was slowly going out of fashion at the time, Kepler made the following argument that Jupiter must be inhabited:

> Our Moon exists for us on Earth, not for the other globes. Those four little moons exist for Jupiter, not for us. Each planet in turn, together with its occupants, is served by its own satellites. From this line of reasoning we deduce with the highest degree of probability that Jupiter is inhabited.[20]

Kepler was a deeply spiritual Christian, inclined to see God's handiwork and purposes everywhere. Nothing for Kepler was just a "fact." God created the world, and his purposes were embedded in the warp and woof of its details, from the number of the planets, to the shape of their orbits, to our glorious moon, to the satellites around Jupiter. It simply could not be that the moons around Jupiter were just random orbs, attached by gravity to another random orb. And what purpose might those orbs serve? They must be "lesser lights to rule the night" for the good citizens of Jupiter—the Jovians—just as Earth's moon was a lesser light to be enjoyed by the citizens of Earth.

In the years since Kepler spoke with such naive confidence about the Jovians, interest in extraterrestrial life has remained high. Emanuel Swedenborg, born a half century after Kepler died, claimed to have conversed with life-forms from other planets. A staunch Lutheran, like Kepler, the well-educated Swedenborg experienced all manner of religiously oriented visions, including conversations with spirits from Jupiter, Mars, Mercury, Saturn, Venus, and even the moon. Science-fiction authors from H. G. Wells and Jules Verne to the scriptwriters for *Star Wars* and *Star Trek* have populated the universe with diverse life-forms. The result has been a grand menagerie of Selenites, Vulcans, Klingons, Borg, Wookiees, and Ewoks.

Speculation about alien life-forms has long been a source of great entertainment and even much provocative science fiction. Sometimes aliens are benign, as in the episode of *Bewitched* pictured above where the confused Aunt Clara summoned an alien by mistake. More typical is the threatening alien life-form such as that developed in H. G. Wells's classic *War of the Worlds*. This image is from a 1906 French version of the book.[21]

The most popular speculation has been Martians, immortalized in H. G. Wells's classic, *The War of the Worlds*, which had Martians invading Earth. Mars is a natural choice for such speculations since it is the most similar to Earth. Mars is also close enough to study in detail. In the late nineteenth century, the Harvard-educated Percival Lowell trained his telescope on Mars and became convinced he saw artificially constructed canals. Letting his imagination run wild, he inferred that Martian engineers had built a vast network of canals, which were on such a scale that the entire planet must have cooperated on the project. The Martians, he reasoned, must be a peace-loving population.

More careful observations turned Lowell's system of canals into a mirage, and almost all scientific evidence for life on Mars has since completely disappeared. In 1965 the *Mariner 4* flew close enough to Mars to take good pictures of the surface. At the time, scientists were hopeful that the mission would turn up evidence for life; in the best of all possible scenarios a friendly Martian would be caught on camera, waving at the passing space probe. The Martian surface, however, proved to be dry and barren; the atmosphere was too thin to enable any water on the surface; and no friendly Martians waved hello as the *Mariner* sailed silently past, its camera feverishly shooting photos and transmitting them back to Earth.

The prospects for interesting life on Mars turned out to be as dry and barren as its parched surface. Later missions looked more carefully for biological activity in the Martian soil, and Martian meteorites that have made their way to Earth have been scrutinized for evidence of life or its byproducts. All of this has turned up nothing, beyond the occasional molecule that some eager biologists think might have been produced by a simple life-form.

As the prospects for finding life next door in our solar system were winding down, another project to find life far away was ramping up. For decades scientists have understood that there are many stars like our sun, and thus have every reason to suppose that Earth-like planets might orbit about them. Maybe, if we can't go visit our extraterrestrial neighbors we can at least call them on the phone.

In the early 1960s a meeting was convened in Green Bank, West Virginia, to start thinking, intelligently if possible, about how many extraterrestrial civilizations there might be outside our solar system. Were there any grounds for optimism that life might be found "out there"? The astronomer in charge of the discussion was Frank Drake, who pondered this question and wondered exactly how one might estimate the probability of other civilizations with which we might communicate via radio signals. With the nearest solar systems being trillions of miles away, any actual voyages were simply out of the question. The best we can hope for is to receive a radio signal from some far-off civilization. The search that intrigued Drake and the others at Green Bank was thus for a radio signal that could not, of course, be produced by any random life-form. (Frogs, for example, can't build radios.) Only a technologically advanced civilization that was capable of broadcasting radio waves into space would be detectable from Earth.

Drake's problem was challenging. The universe is really big. How many civilizations are out there to be found? And are they close enough that we could get a signal from them? It takes years for a radio signal to reach Earth from the *nearest* star. Most stars are so far away it would take millions, and even billions, of years for a signal to reach us. Coming up with an estimate seems like a losing proposition. Nevertheless, it was foolhardy to spend money listening for a message from another civilization if there was no reason to believe one was out there. So Drake worked up an equation of sorts.

Drake's equation is simple—just a bunch of numbers multiplied together. But the numbers are tricky to interpret and complicated to estimate for various reasons. For example, a civilization might come into existence, broadcast radio waves for a million years, and then vanish when its sun burns out. Or they might destroy themselves in a war, or a plague might kill everyone. We can never make contact with a civilization that disappeared a thousand years ago, or one that will appear a million years from now. In similar fashion, our civilization produced no radio waves until recently. If another civilization was "listening" to Earth at the time of Newton, they would have picked up no signal, and thus no evidence that there was life here.

The Drake equation, as it is known today, looks like this, where N is the number of civilizations with which we *might* possibly communicate, via radio technology:

$$N = R^{*} f_p n_e f_l f_i f_c f_l$$

The terms on the right-hand side are defined as follows:

R*= The rate of formation of suitable stars such as our sun. Since stars, like people, are born and die, we can't simply count the number at any one time. William Shakespeare and J. K. Rowling cannot communicate with each other since their lives had no overlap.

f_p = The fraction of those stars with planets. Not every star has planets, but evidence suggests that planetary systems may be common for stars like our sun.

n_e = The number of Earth-like worlds in a planetary system. By "Earth-like" we usually mean having liquid water on the surface. There is a restricted habitable zone around a star where the temperature is in the range where water will be in liquid form. In our solar system, Venus is too hot and Mars is too cold. Only Earth can have liquid water.

f_l = The fraction of those Earth-like planets where life appears. Life is complicated and may not necessarily appear just because the conditions are right. We don't know how, when, or where life originated on Earth so it is hard to estimate how common life would be, even on Earth-like planets. But we have to make some intelligent guess.

f_i = The fraction of planets where life evolves to produce intelligent creatures. If life appears but remains forever at the single-cell level of complexity, then there will be no intelligence. In fact, life has to produce some rather brilliant organisms before radio technology becomes likely.

f_c = The fraction of planets on which electromagnetic communications technology develops. Life needs to be intelligent, of course, to communicate, but it requires the additional step of becoming technologically advanced to communicate via radio waves. It is an extraordinary thing to develop languages and to speak to each other in the highly symbolic way that we do. It is quite another to broadcast a language using radio waves. Humans were speaking for thousands of years before anyone even knew what radio waves were, much less how to communicate with them. There are also some interesting questions related to our physical capabilities and what kind of manual dexterity, in addition to intelligence, is necessary to create technology. Could dolphins, for example, create technology?

f_l = The percentage of the "lifetime" of the planet that civilizations exist with communications technology. This is perhaps the most speculative part of the equation. Our own experience is that knowledge of how to use radio technology arrived billions of years after the planet was formed and about halfway through the lifetime of our sun. It also arrived in the same set of ideas as the knowledge of how to build bombs that can render our planet uninhabitable. With the cold war now over, it seems reasonable to speculate that we will not destroy all life on this planet and it will persist for the remainder of its lifetime. But who knows? And who knows what alien civilizations might do with nuclear technology?

Most people find the Drake equation strange and unreliable. It is, ultimately, little more than a series of educated guesses multiplied together, with each component making the whole package ever more speculative and uncertain. But it is the best we have for thinking "scientifically" about communicating with life on other planets.

The Drake equation is typically approached with great optimism by people who really want to find other life in the universe. Optimists assume that *every* planetary system about an appropriate star will have an Earth-like planet. Life will emerge and acquire both intelligence and technology on *every* one of these planets. And this technology will *never* lead to the destruction of the civilization that produced it.

These may be reasonable guesses. Or not. On our own planet, we have the interesting possibilities that dolphins have developed a high level of intelligence but no technology. It could very well be that intelligence must be coupled to a high level of manual dexterity to create technology. Dolphins, for example, are physically unable to solder two wires together on a circuit board. No matter how great their intelligence or how many years we gave them, it seems unlikely that they will discover electricity and use it to create radio technology.

The optimistic result from the Drake equation leads to the suggestion that there may be as many as 10,000 communicative civilizations in the Milky Way. This may seem like a lot but the Milky Way is so huge that many of these, even if they exist, will be so far away that we could not realistically expect to communicate with them. There are stars on the other side of the galaxy from us that are more than 50,000 light-years away. The radio waves that we began producing a few decades ago will not reach them until the year 52,000. Similarly, if they started producing radio waves even 10,000 years ago, we won't know about it for a very long time.

The situation gets even more discouraging when we think about what it might mean to actually "communicate" with an alien civilization. Suppose we did get a signal from an Earth-like planet in orbit around a star in our part of the Milky Way galaxy. Even the close stars are so distant it would take decades for such a signal to arrive at Earth. Imagine "talking" to someone who took fifty years to respond. "Hello," you say now as a college student working in the Green Bank laboratory. "Hello. Who is this?" comes back fifty years later, when you are living in a retirement community.

To date, no signal of any sort has arrived to reward the optimistic astronomers who don't want to be alone in the cosmos. They have been listening in vain for a half century. And, if the Drake equation tells the full story, there may be no signal any time soon, if ever. More recently, some Earth-like planets have been identified outside our solar system. The Gliese 581 system has some planets in orbit about a suitable star. Three of them have been suggested as candidates for life, but one labeled Gliese 581g is the leading possibility. But there is nothing we can do except train our radio telescopes on that star system and hope for the best. Perhaps happy dolphin-like creatures are cavorting blissfully in warm oceans under the Gliesan sun, oblivious to the possibility of developing radio technology. We will never know.

Four hundred years ago Kepler could happily pronounce that there must be life on Jupiter. Today we look at a very different night sky—one that does not seem likely to be populated with the many creatures of our science-fiction stories. The realistic scientific assessments of today have made us increasingly aware that we may never encounter another species and, while they may certainly exist, for practical purposes they may as well not be there at all. We are like an ancient tribe living on a remote island with no way to travel. There may indeed be many other tribes on other islands but we know nothing of them.

Such considerations motivate a reverence and an appreciation for life. The most remarkable statement that can be made with certainty about our wondrous universe has to be that life exists on planet Earth. What if our planet is the only place in the entire universe teeming with life? Can we reasonably expect there to be life all over the universe, when we know nothing of how it would arise? How exactly did life come to be so omnipresent on our planet? Was the process predictable and straightforward? Or is life a long shot under the best of circumstances?

For a question that is literally so close to home, we know little about how life came about on planet Earth. The process is mysterious enough that some religious believers assert that it takes a literal miracle. The process, they argue, is outside the realm of science and in the hands of God. But it would be strange if the God who worked so carefully with and through natural laws to prepare the universe for life suddenly reversed direction and used supernatural intervention on this one step. Furthermore, we know enough about life to imagine plausible scenarios whereby it may have originated in the distant past. And most scientists believe these scenarios will one day answer the question of how life arose.

Charles Darwin offered one of the most interesting speculations about the origin of life. In an 1871 letter to a friend he suggested that we could "conceive in some warm little pond, with all sorts of ammonia and phosphoric salts, lights, heat, electricity, etc., present that a protein compound was chemically formed ready to undergo still more complex changes." Darwin's suggestion had little scientific basis. He did not know about genes and DNA, so reproduction was a mystery to him. He called his primordial life-form a "protein compound" and its subsequent reproduction and evolution was simply the execution of what he called "complex changes," certainly an understatement in light of what we know now.[22]

In the same year that Darwin wrote this now-famous letter, the greatest physicist of the time, Lord Kelvin, suggested that "the germs of life might have been brought to the earth by some meteorite."[23] Kelvin's speculation that life on Earth has extraterrestrial origins is called *panspermia*. He made the suggestion at a meeting of the prestigious British Association for the Advancement of Science. It was based on the intuition that the early earth was not conducive to the production of life, no matter how simple we imagine that life to be.

Panspermia is hardly a satisfactory explanation for anything. To the question, "How did life arise on Earth?" it simply answers, "It came from somewhere else." It doesn't answer the question at all—it just relocates it. Despite the emptiness of its rhetoric, however, panspermia has had some heavy hitters in its camp. Francis Crick, for example, who won the 1962 Nobel Prize for discovering the structure of DNA, suggested that the earth was "deliberately seeded with life by intelligent aliens."[24] People without Nobel Prizes who say such things are usually viewed with great suspicion, or at least a raised eyebrow. Sometimes men in white coats come to see them. The astronomer Fred Hoyle, whom we met in chapter 1, also argued that life had come from outer space. In his 1983 book, *The*

Intelligent Universe, he made an interesting, if bogus, analogy to the likelihood that nature could create life on her own:

> A junkyard contains all the bits and pieces of a Boeing 747, dismembered and in disarray. A whirlwind happens to blow through the yard. What is the chance that after its passage a fully assembled 747, ready to fly, will be found standing there? So small as to be negligible, even if a tornado were to blow through enough junkyards to fill the whole Universe.[25]

Just as a tornado is not going to assemble a plane from random mechanical parts, so nature is not going to assemble a life-form from chemical parts.

Kelvin, Crick, Hoyle, and other leading scientists join science-fiction writers, prescientific speculators like Kepler, and contemporary creationists in their belief that life on Earth originated in the heavens, although they certainly have different ideas about what "the heavens" means in that context. Perhaps something in the grandeur of life as we understand it today tugs at the soul of the atheist and the believer alike, coaxing them to look up and imagine that something mysterious beyond the night sky is responsible for life on Earth.

Whatever the motivations for looking "up," most scientists today look "back" for an understanding of the origins of life on this planet. Darwin's intuition, like so much of his work, has stood the test of time, awaiting and anticipating scientific progress that would transform his speculations into research projects.

Perhaps the most surprising aspect of life on Earth is how soon it originated. Even the simplest living things leave traces of their existence in rocks that geologists can study to figure out how old they are. Such studies indicate that life arose on Earth about 3.85 billion years ago, not too long after the surface of the earth became solid. (The earth originated as a molten mass and then cooled off gradually until, about 3.9 billion years ago, the surface had cooled enough to become solid and begin to support lakes, oceans, and other water features.) The quick arrival of life suggests to many that it was, as Harvard biologist Stephen Jay Gould once put it, "chemically destined to be."[26]

Predestination is too strong a term, of course, to describe the origin of life in the universe, at least from a scientific perspective. But, just as the finely tuned laws of physics appear configured to produce the chemistry necessary for life, so the structure and behavior of molecules seems to be such as to bring about that life. The problem for science has been the great complexity of the simplest cell and the inability of such cells to fossilize to let us know what they were like when they were young. No scientists, however, envision the process as Hoyle did in his famous tornado caricature. Life certainly did not arise from billions of highly specific molecules just randomly coming together all at once, like a plane assembling in a junkyard as a tornado passes through. Far more likely is a process stretching over millions of years as amino acids, the building blocks of life, became increasingly plentiful in ancient oceans. And then, as the early earth gradually became covered in a blanket of complex amino acids, the construction of ever more complicated structures became more common and straightforward.

This elusive process had to be one where a foundation was laid and other structures were then built on that foundation and then yet other structures were added and so on. It surely started with amino acids—common molecules with a dozen or so atoms arranged in a particular way. Amino acids are plentiful and can seemingly be created anywhere. We find them, for example, in meteorites that fall from the sky, and chemistry students make them in their labs.

A lot of amino acids are needed to make a protein. You have to string 1,055 of them together in exactly the right order to make collagen, for example, an important component in connective tissue, and the most abundant protein in mammals. Collagen is found in tendons, ligaments, and skin. It's in our corneas, bones, and blood vessels. The long string of amino acids required to make collagen has to fold up in exactly the right way to create the solid three-dimensional protein, which then performs important functions.

Hemoglobin is another such protein. It is only 146 amino acids long but the sequence is precise and unlikely to just spontaneously come together by chance. There may be as many as a million different proteins required to keep us running smoothly. The simplest original life-form would have fewer, of course, but certainly many more than would just randomly come together.

Making proteins, however, does not get us very far. The proteins need to copy themselves, and this they cannot do. A pile of proteins, no matter how rich and improbable, will just sit there unless they can be copied. More likely they will disintegrate and go back to being a collection of amino acid fragments. The copying process is carried out by DNA, itself another long string of precisely ordered amino acids. But DNA is more interesting because its structure—the famous double helix discovered by James Watson and Francis Crick—can "unzip," split into two parts, and then each half can rebuild itself from raw materials in the local environment to create two versions of the original. Strands of DNA, with some intermediate steps, can then provide instructions for the assembling of proteins.

The final component is the package—the little bag that contains everything needed for life and that keeps the parts together as an organized unit. Known as the *membrane*, this little bag—itself made of protein—protects the complicated information-bearing molecules inside so they can do their work without being destroyed.

This is the living cell—informational molecules, various proteins, and other structures all happily coordinating their activities inside the protective enclosure of a membrane.

Without all this cooperation, there can be no cell, no life, no possibility for the evolution of complex creatures like us.

The physicist Paul Davies, who likes to work on complex problems in the penumbra of scientific understanding, has written a fascinating book on the origin of life, which he titled *The Fifth Miracle.* The name derives from the miracles presented in the creation story in Genesis, with the origin of life being the fifth miracle. The first four were the creation of the universe, the creation of light, the creation of the firmament, and the creation of dry land. After these four preliminary, foundational miracles, Genesis 1:11 states, “Let the earth put forth vegetation.”[27]

Davies, who does not think for one minute that the origin of life is a traditional supernatural miracle, acknowledges that the process is deeply mysterious. Commenting on the high level of cooperation needed for even simple life, he writes, “If everything needs everything else, how did the community of molecules ever arise in the first place?”[28]

The short answer to this profound question is, alas, we don’t know. We do know how simple life-forms function, and every high school biology book has descriptions of the cell and its many capabilities. There are no magical machines in the cell that resist all efforts to figure out what they do. And even the process by which a DNA strand unzips and copies itself is understood.

The mystery of the origin of life is the mystery of how to cross the barrier from nonlife to life. We can readily explain how the early earth came to be rich in the necessary building blocks of life. And we can see how these building blocks collaborate in a simple living cell. But we cannot see exactly how these blocks managed to arrange themselves into the living, metabolizing, reproducing cell.

Timid thinkers are tempted at this point to throw up their hands and declare that a divine intervention is required. Science will never find any natural process capable of lifting a humble bag of amino acids onto that exalted platform we call life, they claim. Such timidity is not warranted, however, given how much we understand about many of the required steps.

We are in the position of a detective examining a travel document of a suspect who clearly has gone from Boston to Australia. The evidence reveals a flight from Boston to New York followed by a cruise to Bermuda. An appearance a few days later in London suggests the suspect took a second cruise to London, but we don't know which one. An appearance two days later in Australia implies that leg of the trip was by plane, but we don't know the details. The suspect's journey may contain many puzzles, but there is no reason to despair that these puzzles cannot be solved, or to suppose that some kind of mysterious "transporter" technology was used.

The origin of life is a journey with many steps, some of them unknown. We know the beginning and the end, and some of the parts in between, but work still needs to be done.

The origin of life on Earth might be compared to the origin of the Internet. The incredible complexity of the Internet could not have arisen all at once. First there had to be a world filled with increasingly complex electronic components, such as transistors, capacitors, and diodes. From these came devices such as televisions, tape players, and lasers. Simple home computers emerged in the 1970s. Then more complex computers that could be networked became common, and finally we have the Internet. The end result is complex beyond description but, because we know the steps, it is not mysterious.

In one of my favorite movies—the blockbuster *Terminator 2: Judgment Day*—the lead character is a robot from the future played by Arnold Schwarzenegger. As the story unfolds he explains to the puzzled human protagonist how it happened that a vast computer network called Skynet declared war on humanity:

> The Skynet Funding Bill is passed. The system goes on-line August 4th, 1997. Human decisions are removed from strategic defense. Skynet begins to learn at a geometric rate. It becomes self-aware at 2:14 a.m. Eastern time, August 29th. In a panic, they try to pull the plug.

The interesting speculation here is the possibility of a transition to self-awareness that occurs, suddenly, at a specific time. The implication is that when a certain threshold is crossed, brand-new behaviors emerge. Could this be the origin of life—steadily increasing complexity until a critical threshold is crossed?

Many processes exist in nature where a certain "trigger level" is reached and something new happens. A nuclear reactor goes "critical" when the process become self-sustaining. An avalanche can be triggered by adding just a tiny bit more snow. A panic can create a runaway collapse in the stock market. These are all examples of important thresholds—key levels where something interesting happens and the behavior of a large system is dramatically altered.

Almost four billion years ago an important threshold was crossed on planet Earth and collections of chemicals went from simply assembling and disintegrating to *reproducing*. There must have been many intermediate steps, and in time we will figure out the path. Perhaps the path will turn out to be simple and we will strike our palms to our foreheads in dismay that it took us so long. Perhaps the path will only be discerned when we know more science. "Life is so extraordinary," writes Davies in *The Fifth Miracle*, "that it qualifies for the description of an alternative state of matter."[29]

However we come to understand life, the evidence compels us to stand in awe. There is every reason to think that life on Earth may have been the first life in the universe, and perhaps the only life. Life can exist only on planets with liquid water and rich organic chemistry. Such chemically rich planets can form only from the debris of exploded stars, and it takes billions of years for a star to reach this point. It would be hard to imagine a habitable planet appearing in our universe any sooner than Earth. Likewise, it would be hard to imagine life appearing on a habitable planet any sooner than it appeared on Earth. We were quite possibly the first.

The appearance of life on Earth, even in the humble incarnation of the simple cell, was an astonishingly promising development in the universe. Dead, lifeless matter was summoned onto another plane of existence and powerful new forces were unleashed. The remarkable copying process used by life was creative and exploratory. Every life-form, no matter how simple, reproduced itself with a tiny bit of creativity. The laws of chemistry and physics conspired to make reproduction a reliable but not drearily uniform process. Each new generation was slightly different, and in that tiny difference was the possibility of progress. A membrane could be slightly thicker, a metabolism more efficient, reproduction more prolific. Nature explored all the nooks and crannies of possibility and took up residence in many of them.

A universe with life in it is so much different than one without. The creativity that generated the first stars, or the periodic table, or habitable planets, retreats into insignificance when compared to the creativity unleashed when life began. "Life is a game," Mother Teresa famously said; the universe began playing this game on planet Earth—and perhaps elsewhere—

some 3.85 billion years ago. The first life-form had nothing more than potential—no eyes, no ears, no appendages, no mechanisms for motion. But built into the creative possibilities were all those things and more. Who could have imagined that a tiny switch would get thrown and eventually life-forms would be able to see? Who could have imagined that bacteria could grow tiny tails to propel themselves about? The history of life turned out to be one surprise after another. "Life is like a box of chocolates," said the irrepressible Forrest Gump, quoting wisdom from his mother, "you never know what you are going to get."

And there was evening and morning, beginning and ending, of the fourth epoch of creation.

And God saw that it was Good.

DAY 5

The Grand Explosion of Life

Then God said, "Let the waters bring forth living creatures in abundance. Let all manner of living things fill the seas and move onto the land and into the air. Let life find, create, and occupy every imaginable niche, maximizing the variety and diversity of possibilities. And let all creatures be empowered to reproduce and respond creatively to changes in their environments, following the patterns laid down by the Logos."

Our planet looks tiny, fragile, and insignificant against the infinite backdrop of the universe behind it. If we inhabited another planet and the earth was part of our night sky, we would think nothing of its underwhelming presence. Carl Sagan, forever chiding us for hubris, famously called the earth, as we noted in the previous chapter, a "Pale Blue Dot," a label he used as the title for one of his books. The earth floating in space seems but a "dust mote." Just as tiny particles of dust are often visible in our homes, floating in a beam of sunlight, so our tiny planet floats, like a dust speck, in the abyss of space, invisible except for the illumination of the sun, 93 million miles away. The analogy is a good one.

What do we make of the "moteness" of our planet? Is it the signature of our insignificance? Does significance scale with size, so our pale blue dot shrivels to irrelevance in the vastness of the universe? Is it a caution against taking ourselves too seriously—a reminder of the folly of fighting a huge war to secure a tiny bit of territory on a pointless dot? Or does it suggest that we should be attentive to the stewardship of the precious dust mote to which we cling, lest we fall into the bleak abyss of space?

Whatever humility is implied by the speck-like insignificance of our home in space, it is doubly implied by the distribution of life in a thin layer on the surface of the mote. All living things on planet Earth exist in a layer called the *biosphere*. All life is distributed through a kilometer-thick shell on the surface of the earth. The biosphere layer consists of water, soil, and air and is stretched over the half-billion square kilometers of the earth's surface. We don't have to travel too far into space until this thin layer becomes invisible.

E. O. Wilson, the greatest naturalist of our time and a prophetic voice calling us to care for life on our tiny planet, offers the following as a way to appreciate the remarkable fragility of life:

> Imagine yourself on a journey upward from the center of the earth, taken at the pace of a leisurely walk. For the first twelve weeks you travel through furnace-hot

The interior of the earth has well-defined layers, starting with a heavy inner core composed primarily of iron, an outer core, a mantle and upper mantle, and a crust. Life is confined entirely to a thin surface layer we call the biosphere.

rock and magma devoid of life. Three minutes to the surface, five hundred meters to go, you encounter the first organisms, bacteria feeding on nutrients that have filtered into the deep water-bearing strata. You breach the surface and for ten seconds glimpse a dazzling burst of life, tens of thousands of species of microorganisms, plants, and animals within horizontal line of sight. Half a minute later almost all are gone. Two hours later only the faintest traces remain, consisting largely of people in airlines who are filled in turn with colon bacteria.[30]

The entire history of life unfolded in the thin layer that our scientists have labeled the biosphere. Every life-form, from the dinosaur to the dodo, both now extinct, arose, evolved, lived, reproduced, and went extinct within the biosphere. The bat evolved its radar system there. This is where the eagle learned to fly and dolphins to swim. Millions of beavers have built their dams here, creating safe places in the wilderness to raise their families. Every nest ever built by a bird was in this thin region. Every mighty oak was rooted in the soil of the biosphere, which held it secure as it grew tall and mighty, making its grand statement.

All the great battles throughout history were fought in the biosphere. A few short decades ago, Adolf Hitler invaded Germany's neighbors to acquire a tiny bit of additional territory in the thin, fragile biosphere. Millions of people lost their lives in that dispute. In 1982 Argentina invaded the Falkland Islands to take a tiny speck of territory from the British, who then sent a flotilla of ships to the region to get back the tiny islands. In 1989 Chinese students briefly claimed Tiananmen Square as their own, until they were violently dispatched by hostile government forces. All over our pale blue dot, political powers from indigenous tribes to ambitious dictators and superpowers eye tiny bits of land they think they should have. Neighbors take each other to court as they dispute exactly where property lines are supposed to be.

All the world's great culture has been produced in that tiny layer, from Job's anguished laments, to Shakespeare's plays, to J. K. Rowling's beloved Harry Potter books. Baseball and soccer and cross-country skiing happen there. Children play *Monopoly* there. Mozart, the Beatles, and Aretha Franklin all made their music there. Pythagoras drew triangles in the sands of the biosphere. Galileo ground lenses made from that sand and fashioned them into the telescopes he used to discover new things in the heavens. Albert Einstein and Niels Bohr argued for a decade about the nature of the electrons in the biosphere.

The thin layer of life blanketing our pale blue dot is home to so much that is remarkable. And yet so many dismiss it because of its *size*, perhaps the most inconsequential property of just about anything, other than football players and television screens. Why is this? Why do we assume that size is so important that it correlates with relevance? Is it a leftover anxiety from when we were toddlers, feeling insignificant because we were "little" and wanting so much to be bigger? Is it a fear of the vast unknown emptiness of space that surrounds us, as though we are trapped in a big dark room? Does being "little" in a "big" universe put us on edge?

There is a far better way to look at the biosphere, however, than lamenting its admittedly pitiful size relative to the vast universe around it. The biosphere contains unimaginable quantities of information. We just fail to notice it because it is not "visible" in any simple sense. But imagine that information was visible. Imagine that it took up space in such a way that things increased in size as they acquired information. A notebook, for example, would become larger when you wrote things on its pages—at the end of the semester it would be much bigger than it was at the beginning. A book with 100,000 words in it would be twice the size of one with 50,000. In such a world, an iPhone would be a million times larger than a pair of shoes. In fact, iPhones would be larger than the moon. Computers would be larger than jumbo jets; high-tech hearing aids would never fit in your ear. Our heads would grow steadily and rapidly as we learned to walk, speak, read, write, and do mathematics. Adult heads would be thousands of times larger than their infant counterparts. This is what things would look like if information took up space.

Our solar system would be completely different if information took up space. Just the nonhuman creatures on our planet would make Earth huge. Every organism, from a bacterium to an earthworm to an elephant, contains a strand of information-bearing DNA that would require many encyclopedias to write out. A gram of DNA contains as much data as a trillion CDs—not a million, not a billion, but a *trillion.* Add to this all the information that humans generate through language, science, art, and mathematics and we are dealing with mountains—galaxies?—of information. If all this information took up space, then the earth would be bigger than the sun. It would be larger than the entire solar system. It would be larger than our galaxy.

Now suppose we view the solar system through these new information-based lenses, where the size of objects is determined by how much information they possess. In our part of the universe the earth dominates everything. Our solar system contains a tiny sun and some minuscule planets. An itty-bitty moon—barely visible—orbits about a gigantic Earth. The night sky is filled with miniature pinpricks of light that we know come from other tiny stars like our sun.

From this new vantage point our Earth is no longer a pale blue dot. What was once a "dust mote" has been transformed into a dominant and overwhelming body that dwarfs everything around it.

This is not an idle exercise. Information is incredibly important and, while toddlers might whine about being too small and look forward to becoming bigger, they never lament they are short on information; by the time they become adults, however, they will have come to appreciate its relevance. They will know that a library is far more significant than the mountain behind it, regardless of their relative sizes. They will be perfectly content to take advice from a five-foot-tall doctor and ignore the counsel of giant football players.

The story of how our pale blue dot came to be a grand library—perhaps the only one—in the Milky Way galaxy is the story of life on this planet. The story is extraordinary, as simple one-celled forms of life reproduced themselves while exploring every imaginable way to be a living single cell, and then discovered the benefits of cooperation. Cooperation led to multi-cellular life that grew seemingly without bounds. The result was steadily growing complexity and diversity until finally the entire planet was tightly wrapped in a spherical blanket of life-forms. Within this organic world, entire ecosytems flourished as life-forms discovered the advantages of becoming part of larger systems. Within the various interlocking ecosystems was every imaginable kind of life, including a most unique and original species known as *Homo sapiens.*

There are many ways to look at the trajectory of life on our planet. Within the borders of that broad path we encounter the creativity that produced everything from the radar system of the bat to the dolphin's ability to leap gracefully from the water and dive back in without making a splash.

The path from the first simple life-form to the present was meandering and, up close, looks random. Life remained in a simple single-cell mode for hundreds of millions of years and then began an almost frantic branching as species split off in response to changes in their environments. The branching process was driven by haphazard events—droughts, the arrival of predators, alterations to the surface of the earth, and so on. The most famous such event was the extinction of the dinosaurs and many other creatures some seventy million years ago when a gigantic asteroid struck the earth and created so much disruption that the climate changed.

Countless extinctions mark the development of life on Earth. Ninety-nine percent of all the species that have ever lived have gone extinct. If history erected a stone every time a species passed forever from the earth, the biosphere would resemble a vast graveyard. For every interesting species we can encounter in the biosphere there are a hundred that we cannot. In some unfathomable way, both the members of a species and the species itself have finite lifetimes. They appear. They flourish. They die.

But the extinction of species need not trouble us any more than the finite lives of people. A life can have great meaning, despite its inevitable demise. Think of the remarkable humans who have made such a difference in the history of our species—William Shakespeare, Isaac Newton, Martin Luther King Jr., Jane Austen, Bob Dylan—and the mothers that gave them life. They are chapters in our story. A species can also be a part of the grand narrative of life without having to live forever.

Viewed from the appropriate distance, the history of life on our planet looks like an unfolding trajectory moving with a gentle bias in the direction of complexity, intelligence, and meaning. The driving force for change is evolution by natural selection, as Darwin figured out in the nineteenth century. Small differences make some members of a species better adapted to their environment. An encroaching ice age increases the value of thicker fur. Droughts provide an advantage to animals that can store extra water, or to plants with deeper roots. Lighter bones make it easier for birds to fly. Longer beaks let hummingbirds reach into deeper flowers. If the differences enhance reproduction, we say the organism is more fit. They will have more offspring and pass on whatever enhancements they have to the next generation. This continues until all members of the species share the enhancements. Nature is constantly upgrading species with the latest technology, whether it is longer legs or bigger brains.

Much of evolution is driven by competition. The hummingbird with the longer beak can more easily satisfy his own hunger, oblivious to the fact that in so doing he leaves less nourishment for his friends with shorter beaks. Many of the beak-challenged hummingbirds will starve and take their genes for short beaks with them to their graves. Throughout the biosphere, life constantly comes to an end as organisms just don't make it for various reasons. The end is sometimes dramatic, as when the fleet cheetah catches the plodding zebra and brings it down in spectacular fashion. Often no event marks the end, as when a flower grows slowly parched until it dries up and collapses.

The occasional carnage in nature led the nineteenth-century poet Alfred, Lord Tennyson to pen his classic phrase describing "Nature, red in tooth and claw," in his poem "In Memoriam A.H.H." Tennyson lamented that nature seemed to care nothing for either the species or the members of a species. "She cries, 'A thousand types are gone: / I care for nothing, all shall go.'" Man, wrote Tennyson, foolishly trusted that "God was love indeed" and love was "Creation's final law." Nature mocks this misplaced faith:

> Tho' Nature, red in tooth and claw
> With ravine, shriek'd against his creed.

This is the conventional view of nature and her processes—blind, purposeless competition among members of species for limited resources. Brother against brother, sister against sister. The tall and the fast dominating the short and the slow. The smart tricking the stupid. The handsome marginalizing the homely. I can see this out my window in the hardy New England woods that are just starting to don their fall colors. The forest is filled with trees of enormous height that tower over my house. Their great height means that their exalted leaves get all the sun. Their vertically challenged counterparts die in their shade and disintegrate, returning their valuable minerals to the topsoil of the forest floor, to further nourish the very survivors that destroyed them.

By these lights, the biosphere is filled with nature's successful gladiators and they stand, covered in blood, sword in hand, among the vast graveyards of the weaker species they have vanquished to extinction. The successful gladiators are the "fittest," and the process of natural selection would be labeled "survival of the fittest" by one of Darwin's contemporaries.

The phrase "survival of the fittest" conferred a certain approval on the process of evolution, rationalizing the waste and carnage reflected in nature's long history of extinctions. Those who despise evolution for the challenges it poses to traditional beliefs about our origins pose rhetorically charged questions like: "Why would the omnipotent Creator of the universe use such a wasteful (and cruel) process of survival of the fittest (meaning that animals have been ripping each other up over millions of years) to bring about the higher forms of life? This view of 'theistic evolution' goes against God's very nature—and logic itself."[31]

This cleverly worded question has been a part of the creation-evolution controversy for more than a century. But note the rhetoric in the question: "wasteful," "cruel," "ripping each other up." God is depicted as presiding over the process of creation and looking on in approval as animals kill each other in cruel ways, like a sadistic Roman politician delighting in the spectacle of Christians being killed in the forums. The bloodied survivors of the carnage live on to produce offspring, and that is how human beings are created: survival of the fittest.

This all sounds rather dreadful, and it is easy to see why the argument is so effective. But this description is a gross caricature of the evolutionary process.

For starters, no animals that could “rip each other up” even existed for almost all of evolutionary history. The familiar images of dinosaurs killing each other harken to the Mesozoic Era, which began 245 million years ago and ended about 65 million years ago. (This era is *recent* by evolutionary time scales.) Life was evolving on Earth for more than 3 *billion* years before the appearance of even simple creatures that could “rip each other up.” During the first few billion years, life was dominated by rather boring, single-celled life-forms incapable of anything so interesting as ripping up their fellow organisms.

Now consider the classic phrase “survival of the fittest.” In the passage above—penned by Ken Ham, America’s most flamboyant antievolutionist—this process is defined as animals “ripping each other up.” Presumably, we are to infer that the best rippers are the most fit. But this is not how it works at all. The so-called fitness of organisms is not correlated with their ripping skills. Fitness is roughly correlated to the number of offspring. More fit organisms have more offspring and send more of their genes into the next generation. Having large families—whether your offspring are oak trees, brook trout, or humans—is what drives evolution.

Let’s apply these evolutionary principles to me. I have two children; the coauthor of my book *Species of Origins* has none. My relative fitness, as I like to remind my long-term friend and colleague Donald Yerxa, is *two* while his fitness is *zero*—the lowest possible value. The current human gene pool has more of my genes in it than his, but this did not happen because I ripped him up or competed with him in any way. Similarly, if carrot-loving rabbits have more babies than their cousins who prefer broccoli, the rabbit population will end up with a preference for carrots—absolutely no ripping required.

To be sure, there is carnage in nature, but is carnage really the *central theme* of the history of life on our planet? Carnage has almost no meaning when we speak of simple one-celled life-forms, or plants, or animals so simple they cannot experience pain. Competition is nothing more than a metaphor when applied to trees working their way skyward to get more sun, or plants pushing their roots deeper into the earth. *Fitness* has a narrow restricted meaning when applied to ants that live in highly cooperative colonies.

These misleading metaphors, unfortunately, are powerful. They can bewitch us and lead us to see patterns and trajectories in natural history that are not there or, if they are there, are entangled with other patterns that may actually be more dominant. Take our own species, for example. How often we hear laments that we are so violent. The evening news is filled with disturbing tales of murder, war, and even genocide. Such images resonate with the received wisdom that we are a violent and destructive species, programmed by evolution to "rip up" our fellow human beings. And yet, how often do we actually encounter genuine violence?

Suppose you are an anthropologist from Mars, studying the human species with plans to report back on what humans are like. You sit on a bench in a busy part of a shopping mall—a convenient place to encounter humans in their "natural habitat"—and watch carefully to see how *Homo sapiens* interact with each other. What do you see and what don't you see? In all likelihood you do not see *any* acts of violence. Even with sustained observation over several days or weeks you are unlikely to observe a murder or even a fistfight. You may not even encounter a heated disagreement. Nothing that could be described as "ripping up" would occur.

What behaviors would you most likely notice? You would probably see people graciously giving directions to strangers who couldn't find Radio Shack. You would see people attending carefully to children, helping someone in a wheelchair get onto an elevator, or playing peekaboo with toddlers they do not know. You would see people waiting contentedly in lines at the theater. One of the most common interactions you would see would be romantic—teenagers holding hands with each other and engaging in flirtatious behaviors, middle-aged couples having lunch, an elderly person pushing his or her spouse in a wheelchair. You would hear far more laughter than angry screaming. The report you would write for your fellow Martian anthropologists would not describe these earthlings as incorrigibly violent.

Now suppose that the evening news had a regular segment called "Mall Violence" in which they gathered sordid stories from around the world and presented them every night as if this was what happened during a typical day at a typical mall. An uninformed person—especially someone like me, who never goes to malls—might draw a different conclusion about the behavior of *Homo sapiens* in their mall habitats. The conclusion would be completely wrong, however.

The respected biologist Lynn Margulis has suggested—provocatively and in the face of great criticism—that natural history has for too long been a tale told by male biologists.[32] Males, she suggests, are quick to see competition, winners and losers, and exploitation, in settings where those terms have almost no meaning. They nod knowingly when they see violence, and miss the peaceful interactions in the background. The male authors of *The Major Transitions in Evolution* speak of the tiny organelles inside cells, called mitochondria, as "encapsulated slaves," subject to ruthless "metabolic exploitation."[33] This is surely odd language to describe what happened to the mitochondria billions of years ago when they made the transition to living inside cells. One could just as well speak of the harsh conditions imposed on the oregano when it is captured by the measuring spoon and forced to become a part of the pasta sauce.

Robert Wright argues in his provocative book *NonZero: The Logic of Human Destiny* that *cooperation* is the better metaphor to describe the driving force of natural history. While not denying that there is competition, he suggests that organisms often improve their fortunes on natural history's winding path by getting along with each other, not "ripping each other up."

Wright's suggestive term *nonzero* comes from a branch of mathematics called game theory that studies the outcomes of interactions as if they were games that the participants want to win. A *zero-sum* interaction is one in which the gain of one party is equal to the other party's loss. If I steal fifty dollars from you I am richer by the exact amount that you are poorer. The "sum" of the interactions is zero. This is the simplest type of exchange, where the gains and losses cancel each other out. But there are many interactions where both parties come out ahead and everyone is better off after the exchange. Suppose I am a skilled and efficient landscaper with all the best tools who lives in a house that is falling down and you are a similarly equipped carpenter whose landscape consists of crabgrass and overgrown bushes. Suppose also that we both charge fifty dollars per hour for our services. If I landscape your property for two hours while you do carpentry on my house, we will both come out ahead. We make a deal to exchange labor in this way and, even though we both negotiate what will help us the most—a selfish approach—the end result is a mutually beneficial interaction where we were better off to enter into an agreement than to do everything by ourselves. Game theory looks at scenarios like this and projects which sorts of interactions will be most fruitful, whether the "players" are people, genes, countries, species, language groups, or economies.

Every imaginable type of game theory interaction exists in nature. In an interaction between a lion and a zebra, the zebra either gets away or is killed and eaten. The zebra never "wins" in any real sense—it just lives to flee another day. In a fight between a boa constrictor and a warthog, the winner eats the loser in the perfect and most graphic example of the zero-sum game. In contrast, the relationship between humans and their dogs provides greater happiness, health, and longevity for both of them. (Research shows that pet owners live longer, probably because the companionship of the pet reduces stress.) The relationship between bees and flowers is decidedly nonzero. The bees get food from the flowers, pick up pollen in the process, and spread it to other flowers, enabling them to reproduce. A sparrow will chirp out a warning to its flock-mates when it spots a hawk, putting itself at risk but simultaneously greatly reducing the risk to other sparrows unaware of the hawk. A flock of birds that cooperates in this way will be much safer than one where each member looks out only for itself.

Nature is filled with examples of cooperation. In fact, there is so much cooperation in nature that it is hard to resist the conclusion that the dice are loaded to produce cooperation and even outright acts of altruism. An especially provocative interaction in nature was observed by a leading primatologist when a bonobo named Kuni saw a starling hit the glass of her enclosure at the zoo where she lived. The bird was stunned and collapsed. Kuni picked up the bird and carried it to the top of a tree. She spread its wings and launched the bird, which fluttered back to earth and stayed there. So Kuni stood watch over the starling until, near the end of the day, the bird recovered and flew off. It is not clear what benefit Kuni could have received from this act of interspecies kindness. She was clearly motivated to help the bird, but her actions resemble nothing so much as straightforward altruism, informed by an impressive understanding of birds. Kuni was a good Samaritan, in every sense of the term.

Throughout the rich biosphere that blankets the earth we find countless examples of cooperation. We find it in the clearly mindless and mechanical behavior of plants; we find it in the instincts programmed into the birds and the bees; we find it in the plant-animal relationship between flowers and hummingbirds. We find it in the gestures of kindness so often proffered by primates. And we find it in our own species. Where did it come from? Game theory offers a simple explanation for the widespread cooperation found in nature.

Recall that evolution is all about reproduction—the passing on of one's genes to future generations. Organisms that successfully pass on their genes have more of their children—puppies, kittens, seedlings, spawn—in the next generation. When the land becomes parched, a maple tree that withstands drought will live longer than its thirsty siblings. It will have more similarly talented, drought-resistant seedlings than the other maples that struggle to survive in the drought.

Game theory explains in a most remarkable way how cooperation helps with this process. Imagine a single-celled organism that reproduces by repeatedly cloning itself until it is surrounded by clones. Each clone is identical, with the same genes. From the clone's "perspective," it makes no difference what member(s) of the group survive—they are all the same. So, in terms of getting "your" genes into the next generation you may as well help your neighbor as yourself. By "altruistically" sacrificing yourself for your neighbor, you send the same genes—copies of yours—into the next generation as if you selfishly destroyed your neighbor to ensure your survival. By the math of evolution it makes no difference what you do.

Now imagine a swarming colony of such clones spread over a vast territory. Suppose that one cell in this vast swarm experienced a genetic modification that made it more likely to sacrifice itself for its neighbors. Because cells reproduce by cloning, this cell will soon be surrounded by similar "altruistic" cells. When danger strikes, these cells will look out for each other. Sometimes one cell will sacrifice itself to save the others. In such cases only that one cell will be lost to the danger, rather than many more. The lost clone has protected copies of itself, so its altruistic ways continue into the future.

In a vast swarming colony of such organisms, imagine that altruistic cells were blue and selfish cells were red. Initially, the vast swarm is red; suddenly, a small speck of blue grows. Gradually, the blue region expands. Threats from outside enhance the blue region as cells sacrifice themselves for their fellow clones. The selfish regions are devastated by such threats that pick them off one by one. Soon the entire swarm has become blue and every cell has a gene that motivates it to look out for its neighbor.

Fast-forward through many generations. Something happens to a cell that leads it to combine with its sibling cells to create the first multi-cellular life. The cells join together in such a way that their fortunes now depend on the survival of the collective, the most significant development since the origin of life. New multi-cellular organisms come to embody a high level of cooperation as the cells start to specialize. Cells on the outside become optimized to protect the collective against threatening incursions. Interior cells become optimized for reproduction and other functions. Freed from the need to protect themselves, interior cells become very good at what they do.

In time, the outer cells evolve into scales that provide protection for millions of different species of fish in the planet's oceans, lakes, and rivers. When a few species of sea creatures begin their migration onto land, the scales become an armor for snakes and other reptiles. Reptiles will flourish and, for a time, there will be a great age of dinosaurs that will rule portions of the planet.

Genetic changes turn scales into feathers, and flying reptiles—birds—appear for the first time. The sky becomes their niche as evolution optimizes the remarkable skills that enable them to fly. The birds fly in great flocks and develop skills by which they help each other—from feeding their young, to flying in formations that reduce wind resistance, to chirping out warnings when danger approaches.

The habitats of the earth gradually fill up as the grand narrative of natural history unfolds across the deep time of earth history. An ever-increasing roster of fish and sea creatures flourish in the waters. Various sea creatures spin off daughter populations that evolve into land animals. And land animals sometimes evolve into sea creatures, as happened with the great whales. Otters and beavers live in both worlds and seem to love their hybrid habitats. The forests, deserts, mountains, and grasslands of Earth fill up with specialized animals and plants that discover niches where they can flourish. Mountain goats learn to navigate the rocky terrain of Earth's many hillsides; polar bears lumber about in Arctic solitude, in space they can call their own because nobody wants it. The skies fill with birds that swoop and sing and build nests in the trees.

The biosphere grows ever more complex. The life history of animals and plants entwines their fortunes until they become a grand system of mutually dependent organisms. Their varied lives interact in countless ways. Some must be regularly sacrificed so that others can live. The magnificent lion must bring down the speeding zebra. Little fish are eaten by big fish, and the big fish are grabbed by the brown bear when dinnertime rolls around.

The grand narrative of natural history produced the most remarkable devices along the way—devices of surpassing complexity, able to gather vast quantities of information. The most obvious is the eye. Millions of species have eyes of every imaginable optical configuration, from the camera-type eye being used right now to read this page, to the compound eye of the housefly that can see you coming from every direction. The emergence of eyes ended the most unimaginably deep darkness in the universe. What is sunrise, with no eye to admire it? What are red and blue, straight and crooked, dark and light, in a world without vision? The slow emergence of so many different visual systems over the course of natural history is the awakening of life itself to the beauty of the created order. Soon evolution would build a rich aesthetic into the patterns of life and reproduction. The choosy peahen would admire the tails of the strutting peacocks until she found one that seemed just right.

Auditory systems emerged and the world shook off its billion-year silence. Waves on the shore, wind in the trees, and thunder on the plains were all heard for the first time. Birds learned to sing, cats to purr, and dogs to bark. The lonely loon began to roll its lament across wilderness lakes, like the one I vacation at every summer in maritime Canada. Higher primates learned to laugh and have fun with each other. Information was shared in all these communications, and many species developed vocabularies enabling them to communicate complicated messages.

Systems of touch, taste, and smell brought other experiences into existence for the first time. Animals learned to delight in the flavors of the foods they needed the most. The sense of smell found many different uses in various species. Ants learned to lay chemical trails that their comrades could follow by smelling them. Males of various species—dogs, foxes—learned the distinctive smell of the female that announced she was ready to mate and create new life. Flowers everywhere developed fragrances that attracted various animals to them. Animals learned to avoid things with offensive smells. Sugar molecules, composed of simple atoms forged in the interiors of stars, tasted good. Animals everywhere came to enjoy the experience of eating sugar.

The world came to appreciate itself in new ways.

It was an amazing process, how the world came to appreciate itself. Rich and wonderful experiences, made possible by the complex machinery developed by evolution and driven in large part by cooperation, became a routine part of life.

And there was evening and morning, beginning and ending, of the fifth epoch of creation.

And God saw that it was Good.

DAY 6

A Child Is Born

Then God said, "Let life grow steadily in its capacity to understand and experience the world at the deepest level and to discover the nature and origins of its own existence. And let life develop its own creative capacities so that the world may be filled with great novelty and rich experiences for all its creatures, following the patterns laid down by the Logos."

Newborn humans enter the world helpless and incompetent. The probability that any of us would live to our first birthday without great investments of time and energy by our caregivers is zero. Even with effort the probability has been disturbingly low until recently. Consider the shocking statistics reported at an orphanage or "foundling home" called the Hospital of the Innocents in Florence, Italy. The home took in more than 15,000 babies between 1755 and 1773. *Two-thirds of them died before their first birthday*, despite receiving basic nutritional and medical care, as understood at the time. At St. Mary's Asylum for Widows, Foundlings, and Infants in Buffalo, New York, the numbers weren't much better. Between 1862 and 1875 more than half of the 2,114 children taken in died within a year, despite receiving "every possible care and attention that the Sisters would allow as to food, ventilation, cleanliness, etc."[34] Even in prosperous London there was a 20 percent chance that infants would not see their first birthday. The numbers were much worse in other cities, such as Manchester and Leeds.[35] In an era when families had many children, most family burial plots had tiny gravestones—sometimes several—marking the deaths of babies. Newborns arrived with both anticipation and dread.

Our literature and art are filled with recognition of the basic fact of infant and childhood mortality, an acknowledgment of a once-pervasive reality that needed to be processed. My mother loved to quote Eugene Field's sentimental nineteenth-century poem about the death of "Little Boy Blue," who "awakened" to an "angel song." The death of the little boy is hidden so gently in the lyrics that I did not realize until I was an adult what the poem was about.

The little toy dog is covered with dust,
But sturdy and staunch he stands;
The little toy soldier is red with rust,
And his musket moulds in his hands.
Time was when the little toy dog was new,
And the soldier was passing fair;
And that was the time when our Little Boy Blue
Kissed them and put them there.

"Now don't you go till I come," he said,
"And don't you make any noise!"
So, toddling off to his trundle bed,
He dreamt of the pretty toys;
And, as he was dreaming, an angel song
Awakened our Little Boy Blue—
Oh! the years are many, the years are long,
But the little toy friends are true!

Ay, faithful to Little Boy Blue they stand,
Each in the same old place—
Awaiting the touch of a little hand,
The smile of a little face;
And they wonder, as waiting the long years through
In the dust of that little chair,
What has become of our Little Boy Blue,
Since he kissed them and put them there.

We enter the world in desperate need of heroic caregivers, after a dangerous birth process. Our skill set to navigate our unfamiliar new world contains little more than an instinct to suckle. Our immune systems are underdeveloped and we are easily struck down by illnesses. We cannot feed ourselves or even find food, unless our heads are conveniently lying on our mother's breast. How did this strange state of affairs come to be? Why are we born so helpless and needy in ways that the

To the Unknown Land by Edmund Blair Leighton, painted in 1911, is typical of many works of art illustrating the all-too-common death of young children. As in "Little Boy Blue," the child is pictured as being taken away by angels, mitigating the pain of the family's loss.

newborns of other species are not? Why can a newborn colt totter onto its feet and be running with its mother within hours of its birth, while we need a year to rise unsteadily onto our feet and totter about? Newly hatched baby birds are often out of their nests in a couple of weeks and can fly not long after. In some species of snakes, the parents are long gone even before the eggs hatch. The baby snakes never even see their parents.

To a greater degree than any other species, we are born prematurely, literally years before we can face the smallest challenges to our survival. We are the result of a Faustian bargain struck by nature—a bargain that shapes much of who we are and the cultures we create. The bargain has to do with our brains—powerfully impressive, but requiring overly large heads to house them. Our heads are so large that we can just barely squeeze through the birth canals that bring us into the world.

There is tremendous value in our big brains—no doubt about it. Nature has worked hard to provide us with prodigious gray matter and we can do incredible things as a result—from writing symphonies, to developing scientific theories, to creating iPhones. On the other hand—and this is the Faustian part—we have the challenge of getting those big brains out of their cozy wombs and into the world. The compromise negotiated with nature as a part of this bargain is that we will be born prematurely, before our brains are fully formed and while our heads are still soft and pliable. We will enter the world long before we are ready in exchange for the intellectual power to understand and transform that world.

The consequences of this bargain are quite surprising. On the one hand, human babies are helpless and vulnerable. The number that died over the millennia represents an unimaginable tragedy, especially as we consider the enduring grief of their families. On the other hand, the helplessness of babies has transformed our species into passionate caregivers. Because babies need so much help to survive in their new worlds, nature has provided our species with instincts up to the task.

Evidence for this is all around us. Who has not been in a gathering of adults fawning over a new baby? What celebrations compare to birthdays? Pregnant women are treated with great respect and affection. Announcements of pregnancies are met with great enthusiasm. Miscarriages are occasions of great sympathy. Curiously, we feel an unusual license to place our hands on the stomach of a pregnant woman—a familial, even intimate gesture that would be completely inappropriate if the woman was not pregnant. It is *natural*, in the best sense of that word, to love and care about babies, even those that are not ours or have not yet been born.

This makes sense in the light of our history as a species. Recall that evolution is all about getting one's genes into the next generation. Your personal, even idiosyncratic, traits will become more common if you have children who carry the genes that make you that way into the next generation. Imagine parents with genes predisposing them to provide unusual attention to their offspring—more love, more care, a greater sense of responsibility. The children of those parents enter the human race with a lead over the children next door, born to parents without such genes. Attentive and loving parents are rewarded by watching their children grow up to have babies of their own—grandchildren, as we call them now. My Aunt Norma has twenty-two grandchildren and twenty-five great-grandchildren and every one of them carries some of her genes. Parents less concerned about the helplessness of their offspring—for many species, not just humans—were not so rewarded throughout evolutionary history.

A baby that survived a challenging childhood because her parents—primarily her mother—had an enhanced "maternal instinct" can pass that instinct on to the next generation. Over time, such caregiving instincts and intuitions became a universal part of parenting in many species. They are now near-universal ingredients in human nature. Remarkably, the instinct often goes far beyond parenting one's own offspring. Parents typically have no trouble nurturing adopted children, and most adults are attentive to the needs of babies they know nothing about. A random commuter once stopped on the road in front of my house to check on my baby daughter who was playing dangerously close to the passing cars. (I had briefly disappeared to fetch a tool.)

Over the millennia we learned to nurture and love the newborns of our species. This love for children is now a defining characteristic of our species and at the heart of the family unit. When a baby arrives in a typical family there is great celebration, and even siblings inclined to jealousy join in the care of the new family member, eventually coming to love her or him.

The love lavished on babies in our species is, paradoxically, connected to our dangerously large heads. Nature, we might say, solved the problem of the big brain by creating big love. Absent this love, no human infants would ever have survived and our species would be extinct.

Brain size and care of offspring are correlated in interesting ways. Fish and snakes have such tiny brains that there is hardly a bulge in their streamlined bodies where the brain is located. They also pay no attention to their offspring. Cows and horses have modest heads relative to their bodies and pay some attention to their young. Primates have large heads compared to their bodies and pay a lot of attention to their young. And finally, the most oversized heads of any species belong to us, and we pay the most attention to our offspring.

Oversized heads are a large part of what make toddlers so cute. As soon as they start walking they resemble little adults with gigantic heads, and nature has taught us to be attentive to tiny adults with large heads. Tellingly, we find any creatures with oversized heads cute—an instinctual clue that triggers our care and affection. Popular and enduring icons such as Mickey Mouse, Charlie Brown, and SpongeBob SquarePants are intentionally drawn with the exaggerated proportions of a toddler.

Our big brains make us smart and able to figure things out. One of the great achievements—perhaps the greatest—has been to eliminate most of the challenges of birth created by our big brains and the challenges of staying healthy while our immune systems mature. This has been done so effectively that infant mortality has been almost eliminated by modern medicine, wherever it is available. The same is true for childhood illnesses. Raising children is one of the most meaningful activities for our species, and the expectation of new life in a family is no longer accompanied by a sense of foreboding.

The remarkable power of our big brains defines our species. We cannot see like the hawk, swim like the dolphin, run like the cheetah, or even detect smells like our dog. We have no radar like bats. We cannot spray venom on our enemies like skunks, or poison them like snakes. Our only truly remarkable skill is *thinking*, made possible by brains with far more power than they need to simply help us survive. The story of how we got these big brains is the story of how we came to be human.

We share our planet with lots of creatures; some have brains and some do not. Some have structures that may or may not be properly called brains. The simplest life-forms don't have brains, even though they are often observed to do "smart" things. A houseplant, for example, leans toward the window to get more light, just as a person might walk to a window in the evening to read something in the dim light of sunset. Both the plant and the person need light so they move toward the window. But the plant does not think to itself, "Photosynthesis will work better with more light, so I will lean toward the window." The plant will lean toward the light and right into a candle flame if one is in the way. The plant's action is entirely mechanical, like a ball rolling down a hill. Nature provided plants with an ingenious mechanism to steer them toward light. A growth hormone called auxin in plants is repelled by light so it moves to the darker side of plant stems and trunks. By stimulating excess growth on the dark side, the auxin bends the plant toward the light. Countless life-forms have similar mechanisms that produce beneficial behaviors that look intelligent.

Primitive brains appeared when nature began to *concentrate* intelligent activities—meaning sensory interactions with the external world—in one location. The process was so gradual, however, that there is no clear point in history when animals sporting brains appeared. The humble flatworm—a group that includes the infamous tapeworm that can live in human intestines—has a "head" containing multiple sense organs that do some of the same things that we do with our heads. But the flatworm could hardly be described as intelligent in any meaningful sense.

Nature built upon good ideas. Brains continued to grow until, by the time reptiles dominated the planet, brains had become complicated control centers. Functions involving the coordination of multiple body parts—getting a "hand" to a location specified by the "eye"—were conveniently located in the brain, around which protective skulls developed. Inside those skulls sensory neurons communicated with the rest of the body, keeping hearts beating, lungs breathing, body temperatures constant, and bodies in balance as they moved.

A vestige of this "reptilian command center" remains tucked away inside our head today, still performing the same functions under the radar of our consciousness. This is the oldest and most basic part of our brain, and it sits atop our spines like the handle of a baseball bat stuck in the middle of our heads.

The functions performed by this ancient command center are taken for granted and even dismissed as trivial. After all, salamanders can breathe and balance, so how hard can it be? But nothing could be further from the truth. The skills required for balance are complex. After decades of work and millions of research dollars, our most brilliant computer scientists still cannot build a robot that comes close to balancing itself like we do. The computing power required to maintain balance while moving over irregular terrain is prodigious. Picture a child leaping from rock to rock while crossing a stream, each successive perch different from the last. That our brains do this on autopilot without us having to "think about it" is one of the great blessings of our construction. The cheering that accompanies a toddler's first halting steps is entirely justified, for walking is a major achievement.

Wrapped around the ancient command center that we inherited from earlier, pre-mammalian life-forms is the limbic system—the first brain unit unique to mammals. Our emotions live here. The limbic system stores memories of our experiences—the contentment we felt on mother's lap and the humiliation of playground teasing, the great stimulation of desserts and the unpleasant smell of rotten eggs, the thrill of romantic attraction and the frustration of rejection, the satisfaction of reading a good book in a comfortable chair and the disappointment that the movie was not as good. All this emotional data is stored in the limbic system, to be recalled when needed.

The limbic system uses these memories to guide our decision making. We seek our mother's comfort when we are troubled because we remember how nice it was just yesterday when we needed that comforting. We avoid the playground bully because we still wince from the pain of his taunts last week. The familiar image of the bully in our visual system summons the relevant emotion from the limbic system. Each new experience is less new, less dangerous, less ambiguous, as stored memories remind us of what it was like last time. Like the more ancient control center nestled underneath it, our limbic system exerts most of its influence under the radar of our consciousness. We recoil in a most natural way from unpleasant experiences, unaware of exactly why we did so.

The most human part of our brain—where the cool stuff happens—is the neocortex. This is what comes to mind when we picture our "brains"—the pair of mottled hemispheres resting side by side like two lumps of bread dough ready to go into the oven. Here is the seat of what really makes us human—our capacity for language, abstract thought, art, music, imagination, worship, moral reflection. All of the culture on our planet resides in this part of the brain, a little bit in each one of our seven billion heads.

The three parts of the brain are not sealed off from each other, of course. An incredible number of communication links makes it possible for the different parts to talk to each other. A starving person can make themselves eat food that, on a better day, they would avoid. The rational message from the neocortex—you need to eat or you will die—trumps the more visceral response from the limbic system—stay away from this foul-smelling stuff. A deeply moral person can, with effort and commitment, forgive an enemy responsible for negative experiences catalogued in the limbic system. A sudden romantic stimulus can even overrule the old reptilian control center and increase our heart rate. Some remarkable individuals can, through the force of will summoned in the neocortex, slow their heart rates or lower their body temperatures. The brain, when all is said and done, is a remarkably coherent organ. It is, after all, what we think of as our "self."

Our brains started expanding about six million years ago, when our human-like ancestors first started to walk upright. Walking upright had advantages, such as freeing our hands for other things—picking fruit or carrying tools. But to become really good at walking required many other changes as well. Creatures that walk need hips that are close to each other so their center of gravity doesn't go back and forth too much as they move. Knees need upgrades if just two of them are going to support the full weight of the body. And, most important, brains need to be bigger in order to rapidly process the increasing avalanche of information necessary to maintain balance. Running on two legs means regularly supporting—and balancing—the body on just one point, and that requires lots of computing power. Enhanced brainpower is as important as stronger muscles and bones. Individuals with extra gray matter had an advantage.

Brainpower differs from bone and muscle in being far less specific in what it does and how it does it. While stronger muscles are incredibly important in pulling two bones together—think biceps doing arm curls with dumbbells—they don't make much contribution in other areas. Stronger biceps contribute little to one's running speed or one's ability to kick a soccer ball. But brainpower developed by nature to manage the complex task of moving on two legs contributes in countless other areas.

About 2.5 million years ago our ancestors began using their growing brainpower to make tools. In so doing they began the long process of transforming the environment around them for their own purposes. Sharpened stones were used to cut animal hides, remove meat from bones, and even detach large bones from carcasses so a portion of the "kill" could be carried to other locations, such as back to their families. Stone tools could sharpen sticks, useful for digging edible plants. Round stones could crush nuts and open bones to get at the nutritious marrow inside. The resulting improvements in diet worked synergistically with the growing brains of these first toolmakers. Brains demand a rich and steady diet of calories that was now easier to come by. Brainpower that enabled these improved ways of life was immediately rewarded with improved nutrition.

Improvements in upright posture, longer legs, and general thinking skills made travel more feasible. By 1.8 million years ago our ancestors were traveling to new regions, adapting to new climates, mastering new challenges. Some of our ancient African ancestors traveled as far away as China. By 1 million years ago they were in Britain. Body types adapted to local conditions. In cold climates bodies developed to be short and stocky with layers of fat to preserve heat. In hot climates they were tall and thin, maximizing the area of skin available to release heat. The same body types persist today in Eskimos and Ethiopians, respectively. Powerful, flexible brains enabled these challenging migrations, developing original strategies in response to new challenges.

Our ancestors solved many interesting problems as they altered the world around them. Climate changes required new ways of life even for groups that were not migrating. Streams and lakes dried up and then returned. Vegetation patterns changed in response. The role of the social group—what would one day be called the *tribe*—grew steadily in importance. Tasks that could not be solved by individuals—such as bringing down huge, powerful animals—were readily accomplished by groups. The value of sharing became more apparent. The importance of living successfully in community placed further demands on growing brains to develop powerful new social skills. An ability to recognize a larger number of individuals and understand relationships in the group became important. Being able to detect deception, compassion, fear, loyalty, and anger in the faces of those around you became critically important.

Five hundred thousand years ago, brains experienced a puzzling growth spurt. It was probably driven by the increased role that enhanced brainpower was starting to play in survival, flourishing, and reproducing. We know that intelligent parents must have thrived to produce more offspring. By 250,000 years ago symbolic communication, the precursor to our linguistic skills, appeared. The use of symbols represented a giant leap forward in the ability to communicate. Language could be used to provide descriptions of internal mental states—happiness, anticipation, sorrow—enabling interpersonal and social understanding to reach an entirely new level.

Not long after the explosion in both language and brain size the most significant transformation on the planet since the origin of life occurred: a new species that would one day be labeled *Homo sapiens* appeared. Initially the population was small, not larger than a small city, and not remarkable in terms of its planetary significance. All that would change, however, as these first humans began the long process of transforming their planetary home. Sixty thousand years ago our species began a worldwide migration that would eventually take its various tribes into every nook and cranny of the globe, from Australia to Alaska, from China to Chile, from one Pole to the other.

The last ten thousand years have witnessed the transformation of our planet as our species slowly came to dominate everything it contacted except, perhaps, the weather. We learned to domesticate plants and animals; farming began in earnest and we got steadily better at it until today we grow more food than we can consume. We learned that plants need rain so we created artificial rainfall through irrigation, first by carrying water in containers, then by routing water sources to our gardens. We even learned how to modify plants through selective breeding until today our grocery stores are filled with plants we helped create: apples, oranges, walnuts, rice, lima beans, raspberries, radishes, sweet potatoes.

We bred animals to make them more useful. We turned wolves into dogs that became our best friends; pigs, goats, and cattle were bred from wild ancestors and became staples in our diets. Horses were bred to carry us and pull our plows and wagons. We turned the honeybee into a productive source of delightful syrup.

Our species developed the capacity to imagine worlds that differed from this one and then figure out how to make those imaginary worlds into reality. We now live in a global village and can interact instantaneously with people almost anywhere. We fly in planes and ride in cars. We shower daily and consume natural resources in leisure activities. In so doing we transformed the planet in ways that have begun to raise concerns about long-term sustainability.

At one point in the history of our species we numbered just ten thousand. Every human being on the planet at that time would fit comfortably onto a large cruise ship. Life for those few thousand was incredibly hard and most didn't survive into adulthood. We numbered a few million at the dawn of the agricultural revolution about eight to ten thousand years ago, when we learned how to increase our food supply. By the birth of Christ the world population was over two hundred million. One million of us lived in the prosperous city of Rome alone, enjoying a varied and healthy diet with plenty of fresh water.

Until the nineteenth century, population growth was challenged by great, mysterious, and frightening plagues. Tens of millions of people would inexplicably get sick and perish; pundits and seers would see the wrath of God in the widespread suffering. One of the most devastating plagues, known as the Black Death, killed 30 to 60 percent of Europe's population. The world population declined from 450 million to between 350 and 375 million in 1400.

Bubonic plague didn't completely leave Europe until the nineteenth century, when the world population surpassed one billion for the first time. And then, as the nineteenth century unfolded, the most remarkable achievement of the human race appeared—medical science was born. Infant and child mortality declined rapidly from greater than 30 percent in some rural locations to less than 5 percent. Life expectancy increased. Routine illnesses were routinely cured. Parents no longer feared the imminent death of their children. Poets stopped telling tragic tales of dying children. A great sorrow, long thought to be an essential part of a troubled world, passed into history for those of our species living with modern medicine.

By 1927 there were two billion of us; three billion by 1960; and six billion by the dawn of the third millennium. As I write these words in the spring of 2012 an "Internet population clock" records the world population as 6,998,706,342.[36]

Humans have spread across the planet and transformed the globe. Our prodigious intelligence has made it possible for there to be seven billion of us today and, were it not for political and religious differences that keep us quarreling and fighting with each other, all of us could live in relative comfort. Our influence now even reaches out into space. We are starting to understand our world. Explanations have been discovered

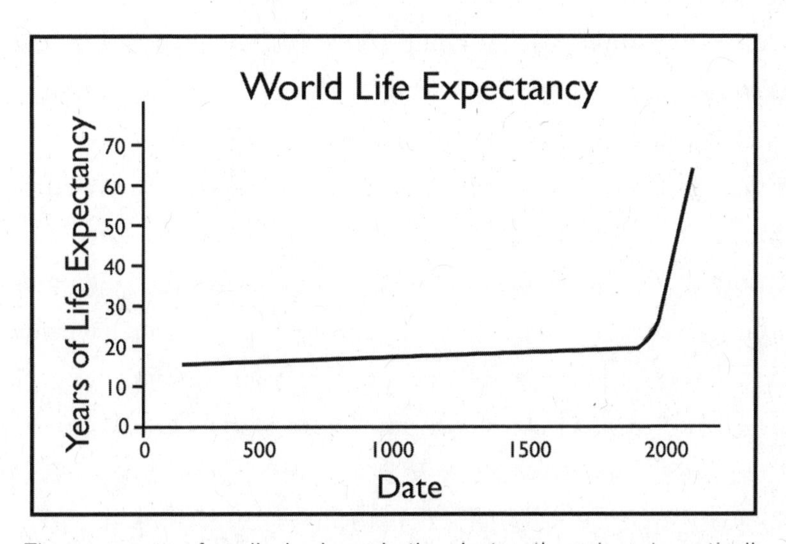

The emergence of medical science in the nineteenth century dramatically increased human life expectancy, largely by reducing infant mortality. It would be hard to imagine a single achievement of our species with an impact comparable to this.

for everything from magnetism to ice ages. We know how the sun shines and why we age. We know the history of our planet, our solar system, and our universe. The cosmos, in a real sense, now "knows itself" because of our species. Human intelligence is the most remarkable feature of a most remarkable world.

Our impressive intellectual achievements are not due simply to the steady growth of our brains, as if size is all that matters. Whales have larger brains than we do, but their extra brainpower is used up keeping their gigantic bodies working properly—it is not available to do calculus, write sonnets, and think about constellations. The important measurement is how big the brain is relative to the body that goes with that brain, as well as the structure of the brain. Ironically, our own intelligence is something that we don't understand very well, although we have learned a lot about how the brain works.

We know what happens in our brains from a purely empirical point of view: electrical connections are constantly being made, broken, and then made again, like a gigantic wall of constantly flipping on/off switches. More electrical signals are generated by one human brain in one day than by all the phones in the world. These signals create seventy thousand thoughts every day—reactions to images and sounds, daydreams, quick calculations, speculations, emotions both positive and negative, anticipations.[37] What we don't understand is how all these electrical connections make thoughts. Imagine, for example, a great wall of electrical connections that could be made in any configuration. You are told that one configuration represents a childhood memory of bread coming out of the oven and another represents a more recent nervous reaction to news that the stock market is crashing. The complex arrangements of connections are clearly different, but there is no way to understand how the arrangements create the two very different mental states. The chasm of understanding between a conscious thought and its associated electrical configuration is vast right now.

We also know that the brain is organized into modules with highly specified functions. There are modules for language, vision, and selecting mates. Babies are born with a "language acquisition device" that lasts for about ten years and then atrophies, which explains why children learn languages so easily but adults find it so hard. Modules for empathy are damaged in the brains of autistic people, making it hard for them to relate to other people. There is even a so-called god module that appears to fire up when people have religious experiences.

Brain modules developed naturally through the course of evolution to solve particular problems, some of them highly specific—such as learning language or balancing on two feet. But the structure of the brain gave it a mysterious flexibility. Each enhancement was like a tool that, while optimized for one function, was so flexible it could easily be trained to do other things. A screwdriver is a useful tool that can drive screws, and most people purchase them for this purpose. But, once you have a screwdriver, you discover all kinds of other things it can do. It works as a small pry bar, or a makeshift knife to slice the packing tape on a box you want to open. In a pinch it could be used as a weapon to ward off an attacker, or it could be pressed into service to stir one's coffee. The handle of a screwdriver can be used as a small hammer to break open walnuts or drive small nails.

The human brain embodies the truth that the whole is often far more than the sum of the parts. The ordinary challenges our ancestors faced in Africa long ago nurtured the growth of complex, multifaceted brains. We no longer face those problems. Instead of evolving to match our environments like our ancestors, we transform our environments to match us. We no longer need a special mental trigger to make us shiver in the cold to avoid hypothermia because we have heaters in our homes. We no longer need to be repulsed by rotting food because we use preservatives and rarely encounter rotting food. We can even adopt children now if we are unsuccessful at finding mates and creating them the old-fashioned way.

But the most remarkable feature of our big brains does not lie in the practical way they can change the world around us. The most amazing of all human capacities is surely our ability to transcend the world of our immediate needs and instincts and *create.* There was never, at any time in the history of our species, the slightest survival value in producing symphonies, and yet we—some of us, not me and probably not you—can do that. The ability of Mozart or Beethoven to imagine a grand work of music, with interlocking and intersecting patterns of melody and harmony and rhythm, to be performed by a tribe of musicians and a chorus of singers, using instruments crafted from natural materials and voices that evolved for talking, is simply inexplicable. It's like a carpenter gathering some pieces of wood, selecting a few well-used tools from his toolbox, and building a Pinocchio who comes to life.

Our ability to discover grand mathematical theories goes far beyond the token math skills our ancestors needed to estimate how many coconuts should be gathered for the cookout on Saturday night. Our mathematicians have discovered grand patterns in a world that lies beyond this one—boldly going where no one has gone before, exploring brave new worlds of equations and theorems, where negative numbers can have square roots, motion can occur in ten dimensions, and parallel lines can meet. These worlds are so different from our own that it is hard to imagine how the relevant navigational skills arose.

Over and over again, we have discovered strange harmonies between the mathematical worlds we discover in our heads and the hidden patterns of the world where we live. We learned how to solve differential equations long before we discovered they were the key to understanding atoms. Mathematical patterns, whimsically created and explored in an abstract world disengaged from this one, often turn out to be perfect descriptions of the hidden behaviors of the real world where we live. In ways that remain mysterious, we have discovered that our world is deeply mathematical.

One of our greatest thinkers, the Nobel Laureate Eugene Wigner, expressed this deep mystery in a provocative essay titled "The Unreasonable Effectiveness of Mathematics in the Natural Sciences."[38] Almost all of his colleagues share his amazement. Sir Roger Penrose—one of our greatest living mathematicians—believes there must be a nonphysical "Platonic realm" beyond the "real" world of the senses to account for the mathematical order we find in the world.[39] Einstein was similarly moved by our peculiar ability to understand deep, but highly impractical, truths about our world: "The eternal mystery of the world," he reflected, "is its comprehensibility."[40]

Our brains are built entirely of molecules—the dust of the earth—to solve ordinary problems such as counting coconuts and walking. Their ability to generate *minds* capable of grasping transcendent realities so far beyond our everyday experience remains the deepest mystery of our existence. Poets, who turn language into music, come closest to expressing the strange character of our species:

> What a piece of work is a man! How noble in reason,
> How infinite in faculties,
> in form and moving how express and admirable,
> in action how like an angel, in apprehension how like
> a god!

The beauty of the world, the paragon of animals!
And yet, to me, what is this quintessence of dust?
(William Shakespeare, *Hamlet*)[41]

And there was evening and morning, beginning and ending, of the sixth epoch of creation.

And God saw that it was Good.

THE SEVENTH, FINAL DAY OF CREATION

The Greatest of These Is Love

Then God said, "Let the members of the species Homo sapiens *grow to understand the meaning, power, and significance of love; let them understand the importance of right and wrong. And let them burn with a deep spiritual hunger to know the God that created them and the world they inhabit. Let them begin to understand the mystery of the Logos that lies at the heart of their existence."*

The most meaningful and defining human instinct is love. We are, of course, also creatures who can be creative, violent, jealous, reflective, funny, clever, depressed. But mostly we are creatures who want to live in a world filled with love that we both give and receive. Love is involved in our conception, our birth, our survival of childhood, and our maturation into adults. Love is at the center of our decisions to marry and have children. Growing old without love is one of our greatest fears. Love defines much of who we are—our behaviors, our choices, our dreams, the cultures we create—and is essential to our flourishing in ways we are just starting to understand.

We need many things to flourish. Food and water come to mind, of course, for they are the material foundation of our existence. Existence by itself, however, is the smallest of consolations—a necessary but not sufficient reason to be. Beyond mere existence we actually *need* love. The absence of love is one of the few things that can lead us to end our lives voluntarily. The greatest tragedy in our literature is the story of Romeo and Juliet—lovers who chose to end their lives because they were not allowed to love each other.

Rembrandt's classic painting *The Return of the Prodigal Son* highlights one of the most moving stories of love in all of literature.

Deep intuitions about love have long motivated reflections that we now call religious. One of the New Testament writers admonishes his readers to "love one another; for love is of God, and he who loves is born of God and knows God. He who does not love does not know God; for God is love."[42] The fourteenth-century Islamic poet Hafiz of Persia wrote, "We are people who need to love, because love is the soul's life, love is simply creation's greatest joy."[43] Even the Beatles sang, "All you need is love."

These reflections on the significance of love have sometimes been at odds with our scientific understanding. Our rational skills—so good at understanding the physical world around us—often let us down when we turn them on ourselves. And nowhere have they failed so utterly for so long and with such dire consequences as in understanding the importance of love.

Most behaviors motivated by love are simple commonsense intuitions. Our need for love emerges as a simple natural urge, like our need for water and food. No thought is required. A mother knows to gently place her new baby on her breast without having to read books about nutrition and bonding. The experience is deeply meaningful for both of them and comes naturally. Parents encourage their children to develop friendships with their peers but also to take time to visit their grandmother in the nursing home. We seek out interactions with others because we want to, not because our therapists tell us to. Humans naturally live in communities.

Despite the commonsense character of our intuitions about love, and the emphasis from our religious traditions, only within the past few decades have we come to understand their importance and validity. Our science, unfortunately, has been so often at odds with our common sense that we have come to think nothing of that confrontation. We accept that science progresses by trumping common sense. The moving earth, to take the most famous example, displaced the commonsense notion that the earth was fixed. Twentieth-century ideas about matter overturned the longstanding notion that atoms were like little balls. The theory of relativity forced us to accept odd ideas about the nature of time. A critic once even challenged Albert Einstein for producing theories that contradicted common sense. The great scientist was dismissive: "Common sense," he responded, with some justification, "is the collection of prejudices acquired by age eighteen."[44]

Science continued to trump our commonsense intuitions about everything, from time and space, to electrons, to the history of the universe into the early twentieth century. But nowhere was the challenge from science more dramatic than in our understanding of love. Remarkably, in the midst of the revolutions brought about by physicists, social scientists introduced their own revolution: physical affection, human contact, and even the emotion of love, they claimed with great confidence, were irrelevant to infants. As quantum mechanics was at odds with familiar notions of matter, this new psychology of early childhood ran at right angles to commonsense notions of child rearing.

This misunderstanding in the social sciences was responsible for the death of tens of thousands of children. Unknown to the science of the time was a central "mystery" that we are at last coming to appreciate. Little children need lots of love. They need to be held, hugged, kissed; they need someone to play peekaboo with them and swing them in a circle. Something in these natural, primitive activities strengthens children in mysterious ways, making their immune system more robust, giving them the strength to fight off childhood illnesses.

Disturbing data supports this, such as the example of the Hospital of the Innocents in Florence that we looked at above, where ten thousand babies died in a facility that provided them with basic physical necessities. Despite receiving what was thought to be appropriate care for infants, most did not live to see their first birthday. Such dreadful stories were commonplace before we learned in the mid-twentieth century that love is as essential to health as food and water. Why were so many children dying?

Once germs were discovered and understood, concerned caregivers concluded that they were probably the issue. To prevent deaths from infections spread by touch, the homes developed procedures to reduce the babies' exposure to germs spread by hugging, rocking, or that most ghastly and irresponsible act of germ warfare—kissing. One hospital devised a special box with inlet sleeves that would allow a caregiver to interact with the child—change a diaper, for example—without touching the child. Technicians who handle dangerous chemicals use similar boxes today.

The sterile environments recommended for health reasons must have puzzled and perhaps even horrified some of the caregivers. Instincts to embrace and comfort unhappy babies must have been hard to hold in check. But such restraints fit nicely with the prevailing wisdom in psychology. In the early twentieth century, the president of the American Psychological Society, John Watson, warned of the "Dangers of Too Much Mother Love," insisting that responsible parents refrain from kissing and hugging their children, lest they become emotionally needy, in addition to the risk of germs. Everyone from Bertrand Russell to *Parents* magazine praised Watson's bestseller on raising children.

But still the children kept dying, germs or no germs.

A renegade researcher named Harry Harlow took on Watson and the scientific establishment. In a series of disturbing experiments on primates—experiments that would have gotten him arrested had he done them on humans—he discovered that newborn primates—very similar to newborn humans in many ways—need love and affection just as they need food and water. Primates raised in isolation from parents and peers died sooner and developed serious mental disorders. Socially isolated primate youngsters got sick more often than their socially connected peers. Harlow was convinced the same pattern would hold for humans.

Harlow's groundbreaking work reunited the science of child rearing with common sense. Scientists came to appreciate and ultimately endorse the instincts that lead a mother to carry her baby with her, to hold it close and enjoy physical contact without worrying about germs or emotional dependency.

Across the board, thanks to Harlow and those that followed him, we now understand the deep wisdom of our instincts. We understand that our hearts sometimes know more and know better than our minds can articulate.[45]

The profound love that parents have for children, a love that usually requires sacrificial and altruistic actions, is precisely the kind of love that has consistently been promoted, celebrated, even demanded by the world's great religions. Great religious leaders were teaching the importance of love thousands of years before Harlow showed us that love had health benefits—physical, mental, and spiritual. In ways that we may never understand, our sages, prophets, and mystics grasped something deep about our species long before science.

How do we understand the universal significance of love at so many levels? A hard-nosed materialist might argue, using some of the ideas presented in previous chapters, that love is "nothing but" a survival strategy invented and refined by Mother Nature to facilitate evolution. We love our children because they carry our genes; we love our mates because they are essential for reproduction; we love our tribes because they help us overcome challenges too great to face alone. Love simply helped the evolutionary process along and got built in. It's like fresh air and clean drinking water—things go better when you have some.

Reducing love to an evolutionary tool, however, is a gross caricature of human experience. Something is strangely incomplete about that perspective, like describing a person as "150 pounds of molecules, mostly water." Such a description is absolutely true and would be relevant in some circumstances. But clearly something is missing, making that "explanation" woefully inadequate.

The context in which evolution occurred, as we have seen in earlier chapters, was remarkably effective at enhancing the role of love at many different levels—at building love into the process wherever possible. The religious intuition suggests that this is an important insight into the way things really are, and that love is not merely a survival tool.

As we examine the created order and the many interlocking features of its architecture moving forward from the big bang to the big brain, so to speak, we are moved to ask if there might be a *purpose* to this unfolding history. Momentary snapshots—like our current scientific description of the world—tell us how things are at any moment, but shed little light on where things are going, how they got there, and how the past, present, and future relate to each other. If we line up the snapshots in historical order and look at the sequence, we see change. The question we want to ask is whether the changes that we see suggest any direction or any purpose.

Reflecting on the possibility of such a purpose is a speculative and humbling exercise, at best, but that does not mean we should avoid it. If we want to understand the world at the deepest level we must not be afraid to place speculative and even hopeful explanations "on the table," so to speak, and then step back and ask which perspectives are most illuminating.

We must start with a note of caution. Many of our most brilliant scientists warn that looking for purpose in the universe is a fool's errand, a long-discarded medieval exercise that sensible people abandoned long ago. The Nobel Prize–winning physicist Steven Weinberg, for example, at the end of his classic exposition of the big bang, *The First Three Minutes*, reflected on this quest:

> It is almost irresistible for humans to believe that we have some special relation to the universe, that human life is not just a more-or-less farcical outcome of a chain of accidents reaching back to the first three minutes, but that we were somehow built in from the beginning. . . . It is very hard to realize that this all [i.e., life on Earth] is just a tiny part of an overwhelmingly hostile universe. It is even harder to realize that this present universe has evolved from an unspeakably unfamiliar

> early condition, and faces a future extinction of endless cold or intolerable heat. The more the universe seems comprehensible, the more it also seems pointless.[46]

Weinberg's lament about the pointless universe has gotten a lot of attention, both positive and negative. His observations have a certain authenticity and must be taken seriously. He understands—as most thinking people do—that the world does not present itself to us as a place where a loving God ensures that everyone flourishes. In particular, Weinberg has been touched personally and hurt deeply by the Holocaust, which claimed scores of his relatives. The Holocaust should make anyone suspicious about claims that the world has any kind of happy purpose, if only we look at it in the right way.

Weinberg knows that questions of purpose lead to questions about God, and that questions of God lead to discussions of the problem of evil, and the problem of evil is seen most clearly in the Holocaust. In *Dreams of a Final Theory* he wrote:

> Religious people have grappled for millennia with the theodicy, the problem posed by the existence of suffering in a world that is supposed to be ruled by a good God. They have found ingenious solutions in terms of various supposed divine plans. I will not try to argue with these solutions, much less to add one of my own. Remembrance of the Holocaust leaves me

> unsympathetic to attempts to justify the ways of God to man. If there is a God that has special plans for humans, then He has taken very great pains to hide His concern for us. To me it would seem impolite if not impious to bother such a God with our prayers.[47]

Many scientists of Weinberg's caliber agree with him . . . but many do not. Freeman Dyson, who many believe deserves a Nobel Prize for his work on the foundations of quantum theory, does not see the universe as sterile and without meaning beyond what we invent. He rejects Christian doctrines for the most part, but believes strongly in the importance of religious communities as vehicles for helping the less fortunate. Dyson, whom we encountered in an earlier chapter, thinks that the universe has a purpose, although he is not entirely sure what it is. "The more I examine the universe and the details of its architecture," he writes, "the more evidence I find that the universe in some sense must have known we were coming."[48]

In his Gifford Lectures, *Infinite in All Directions*, Dyson shares his conviction that the universe has a larger purpose, with "evidence of the operations of mind." He speculates that this mind shows up in three places: at the level of atoms, which behave as if they have a modicum of "free will"; our own consciousness; and the "universe as a whole."[49]

"The universe as a whole," he speculates, has "laws of nature that make it hospitable to the growth of mind. I do not make any clear distinction between mind and God. God is what mind becomes when it has passed beyond the scale of our comprehension. He continues, "Our minds may receive inputs equally from atoms and from God."[50]

Sir John Polkinghorne makes a similar argument but from within a traditional Christian understanding. A brilliant particle physicist who shocked his colleagues by leaving physics for the priesthood, Polkinghorne has debated Weinberg—a friend and former colleague—in print and on stage for years. Polkinghorne affirms a traditional understanding of Christianity and delivered a remarkable set of Gifford Lectures in 1993–1994 in which he looked closely at the Nicene Creed—one of the most enduring and ancient summaries of Christian beliefs—through the eyes of a "bottom-up," scientifically oriented Christian. (He uses the term *bottom-up* as a shorthand for an approach that starts with facts and observations about the world and reasons from there to larger statements about purpose and meaning.) Polkinghorne finds, in that oft-cited creed from the fourth century, enduring and relevant truths, but ones that can be meaningfully informed by the latest insights from science.

Polkinghorne is probably the world's leading scholar of science and religion and has been a central figure in that conversation for decades. He has written extensively on the problem of divine action, which is central to any understanding of how God might interact with the world. If we can speak intelligibly about God "creating" a world that has a long evolutionary history like ours, as I have been doing in this book, we must have some idea how God might work within the natural order. Otherwise we are marooned on the desert island of deism, unable to say more than simply, "God started the process long ago," while we sit with our fingers crossed hoping that science won't discover that there never was a beginning.

Does the world offer any hints that God might be able to work within the natural order? Are claims that God might work *through* evolution self-contradictory, because the natural order is so tightly packed with cause and effect that there is no room for anything but disruptive interventions? Polkinghorne is encouraging on this point. He notes that the world is not a system of tightly interlocking parts, but has what he calls an "open grain" that enables it to receive "top-down" input of information.[51] This is a technical, but scientifically legitimate, idea based on well-established ideas from the misnamed "chaos" theory. The basic idea is this: Complicated systems in nature, such as the weather, behave in such a way that they can unfold along different paths, even when they begin at essentially the same point and with the same energy. Such "chaotic" systems, as they are called, can be "steered" with the top-down input of information—instructions that flow to the parts from a global and unified vision for the whole. As the system unfolds it is guided from below and from above. From below are the causal influences from the physical world—energy, forces, masses, and so on. From above are the influences of information—patterns that can be imposed on the systems to push them in one direction rather than another. An analogy here would be the role of laws in determining traffic patterns. The same cars, with the same drivers, on the same roads will do very different things if the law changes. Suppose, for example, that the speed limit was cut in half and drivers were told to use the left lane for driving like they do in England. These "laws" exert a top-down influence on traffic patterns, and very different results could be obtained from the same starting conditions. But traffic patterns would also be affected from below by such things as accidents, decisions to take public transportation instead of one's car, and individual driving habits.

The open grain of the world has to be taken seriously, for it shapes, at the deepest level, the way the world is. This feature of the world is surprising. In the century after Newton, most scientists became convinced that the universe was a giant machine—a clock with tightly interlinked parts connected to each other with deterministic precision. And, just as ordinary clocks had clockmakers, the clockwork universe had a maker—God. But, by the same lights, just as ordinary clocks run without any interaction with their makers, so the universe ran by itself, without any interaction with its maker. There was nothing for God to do, beyond winding up the universal clock "in the beginning." In fact, there was nothing that God *could* do within the natural order, so tightly orchestrated were the unfolding patterns of nature.

The deterministic character of the clockwork universe was disturbing. If everything unfolds in a predictable way from initial conditions set up "in the beginning," then the entire history of the universe was already written at the moment of the big bang. The future is no different than the past—it just hasn't happened yet, like the final chapter of a book you are reading. That chapter is already written and the ending to the story is determined. You just don't know it yet. So, while you may hope that the evil chemistry professor gets his just desserts, you know the result is already determined.

The most disturbing feature of the clockwork universe was its expulsion of free will. If human beings are made of the same atoms and molecules as the rocks and trees—which we are—then we have no more free will than a tree. All this was grim and unsettling, especially to poets and others unimpressed by the cold and clinical picture of humanity provided by science.

The clockwork metaphor for the universe died in the twentieth century as physicists discovered that the universe was nothing like a clock. The processes of nature were open and unpredictable; the present was not determined by the past; the future is not determined by the present; and the final chapter of the story was not yet written and—wonder of wonders—simply could not be written before it occurred. Best of all, humans were not deterministic automatons. Free will, by any measure a mystery, at least was not ruled out by the laws of physics.

In creating space for free will, the open grain of the universe made it possible for human beings to be more than robots following programs specified by the laws of physics. In a real sense, this created the possibility of human action—genuinely free activity that originates with us, as we respond in unique, independent, and undetermined ways to the world around us.

At last we have arrived at the question of God's action. A world with room for genuine human agency also has room for divine agency. God can work within the natural order as a free and independent causal agent, just as we do. As we feed information, top-down, into an unfolding physical world around us, shaping and influencing the course of events, so God feeds information, on a larger scale, into the world.

The possibility of top-down causality uncovered by twentieth-century physics provides, suggests Polkinghorne, a "hint of how it might be that divine providence could also be understood to be at work in history, shaping its unfolding development through the input of some generalized form of information into the open grain of nature."[52]

This model for God's interaction with the world invokes the traditional concept of God's Spirit, which Polkinghorne describes as "actively but hiddenly at work within creation's history." Such notions are speculative, of course, but unless we confine our thinking to a purely materialistic view of the sort that Weinberg found so pointless, we must engage such speculations as a part of the reality we want to understand.

And now we arrive at the heart of our story of creation. If the Spirit of God is everywhere at work in our open-grained universe, that means that every event since the beginning of time has occurred in the presence of God. The history of life on our planet has unfolded with the real option of divine interaction. Events, as they occurred, may have been drawn by God toward fulfillment of divine purposes.

Such possibilities open the door to a different kind of world—one with a real direction to unfolding patterns like the big bang and evolution—and not just in the sense of more complexity or more diversity. If life unfolds in the presence of the Spirit of God, that trajectory may reveal a purpose—a *reason* why the world is as it is.

Such ideas are not scientifically testable. No law relating to divine purpose will ever appear in a science text. But that does not mean that "purpose" is a meaningless concept. We use purpose in a sensible and insightful way all the time when it comes to the actions of free creatures. Free creatures, by definition, are not bound by natural laws to behave in certain ways. My decision to launch myself on my daughters' rope swing in my backyard must be interpreted in an entirely different way than the pendulum-like motion I experience on the swing after I have launched myself. My pendular motion is described by a simple equation that can be solved by high school students; my decision to get on the dangerous swing has no such explanation. In fact, to fully understand my decision to get on the swing would require placing that decision in a larger context and observing my behaviors through time to see what kind of patterns might become apparent when lots of decisions spread through time are compared and connected to each other. The idea of "purpose" would be required to understand my decision.

This is what Freeman Dyson meant when he said, as I quoted earlier, "Somehow the universe knew we were coming." When you connect all the dots across cosmic history from the big bang to the origin of life on Earth they don't look random. They create the unmistakable impression that life like ours is not an accident.

So what do we see when we look across the history of life on our planet? Can those dots be connected in a way that suggests a purpose?

We return to the interesting and multifaceted roles played by love that we have been considering. The challenging circumstances of our birth helped create powerful bonds between parents and their offspring. But this bond is much more than a connection between parents and their children. It forms the basis of a rich network of bonds that connect to society as a whole. Almost everyone shares a concern for babies, toddlers, and children. The powerful family bond extends far beyond the family, seemingly to our entire species in some cases, and even beyond. And, we note, the same is true in other closely related species. Bonobo mothers love their babies passionately also but have been known to extend such compassion to other species as well.

Extended family bonds connect us to people who share our genes. Family reunions are special occasions. As I write these words, my father is enjoying his four new great-grandsons, and there is nothing he likes better than a family gathering where he is surrounded by these younger generations.

What is most interesting, however, about the bonds created so naturally between members of our species, is their flexibility. Parents love their children deeply, even if they are adopted. Although incidents of abuse are higher for stepchildren, they are still incredibly low, and we now understand that humans have no trouble caring deeply for children that are not their own. When my daughters were young I coached their basketball and softball teams. I came to love their teammates as if they were my nieces and I took great satisfaction in watching them learn how to hit a softball or sink a foul shot. My friends that teach young children in the public schools embrace twenty new children every year, and work heroically to nurture them.

As near as we can tell, our ability to love children evolved because it was important that we were motivated to love our biological offspring. But the "programming" that was created was defocused, and extended far beyond our own families.

The rule we seem to be following is, "Care deeply about the young children around you, no matter who they are. Protect, love, and educate them to mature into effective and happy adults."

We have similarly defocused motivations with our extended families or our tribes. On a simplistic level, our extended family is a repository of genes similar to our own. If we think about the grand collection of genes in an extended family, it is not hard to see how those genes might motivate mutual concern for our "tribe." But the motivation is imprecise. We can easily embrace the spouses of our children, nieces, and nephews as though they were genetically related to us. We are not very discriminating when it comes to enlarging the family franchise.

Our shared instincts to love the fellow members of our tribes evolved when we lived in tribes. Our species spent countless generations living a tribal lifestyle in Africa, and various bonding mechanisms emerged to promote tribal cohesion—a cohesion that benefited everyone. We use visual and linguistic clues at a superficial level and shared values at a deeper level to make the relevant connections. But almost any kind of connection can work, which is a bit peculiar when you think about it. Two white adults in a room full of black adults will naturally seek each other out on the basis of nothing other than their appearance. But children of various ethnicities will naturally coalesce around their shared youth and enthusiasm for playmates, regardless of their skin color or even language. They contrast themselves with their adult supervisors, rather than their peers with different color skin. Tourists weary of functioning in a foreign language will be immediately drawn to someone speaking their language, even if they are from another ethnic group. Tribes, both formal and informal, emerge with inexplicable spontaneity, united by everything from love of literature to hatred of the New York Yankees. Tribal unity can be created out of almost anything.

Our tribal boundaries are imprecise. They slide inward and outward and, depending on circumstances, encompass tribes of seemingly any size. We can envision our tribal loyalties as a set of concentric spheres with our most obvious tribe—our immediate family—at the center. As we go out from the bull's-eye we encounter successively larger tribes—our extended family, our church community, our neighborhoods, our town, our state, and even our country. Some of us consider the entire human race as our tribe and there are even those who consider all life to be part of a vast networked planetary "tribe." Science-fiction writers have long speculated that if an alien force attacked the earth the entire human race would set aside their differences and immediately become one large tribe.

Our tribal instincts are tightly interwoven with the moral codes we use to relate to each other. Instincts are not, simply as instincts, moral, although they sometimes appear to be. An instinct can motivate any kind of action, moral, defensive, or practical—birds instinctually build nests, bears instinctually hibernate, and mothers instinctually nurse their newborns. We instinctually recoil from snakes and reach for a falling child. Our instinct to protect a toddler from dangerous traffic—which motivates a behavior that appears moral—is the same sort of motivation as our attraction to sugary foods; neither is—in and of itself—either "right" or "wrong."

Our instincts, however, need not be viewed as the mindless and mechanical productions of natural selection, like our fingers and toes. They developed in the transforming presence of the Spirit of God, incarnating God's purposes. Some instincts are comfortably and reliably transformed into moral statements. Deeply spiritual leaders, able to see past the surface features of the world, are moved by the Spirit of God to articulate the reality of what they believe God is doing in the moral order. These saints call us to rise above the selfish and parochial dimensions of our instincts, to make them universal, to apply them when it does not suit our purposes, to push out the tribal boundaries of our concern until they surround everyone.

Our reflections on the urgings of God's Spirit generate our most important conversations about how we live in the world. Traditions we call religious develop around these conversations as we share together our understanding of how we *should* live in the world, beyond how we *want* to live in the world, or what comes naturally. We come to understand that love must be more than the warm feeling we enjoy when we pick up our own children and they smile at us. We understand that all children deserve this love and, when they are without it, something is wrong in the world—something that we must try to fix even if it means donating our resources.

We have no choice but to care for the helpless—such mandates are as much a part of the created order as gravity and electricity. This is the reality of our religious commitments. We come to understand that we must open the arms of our tribe to everyone who would join us. The unwelcome alien from across the border, the foreigner who does not speak our language, the person who is nothing like us—they belong to our tribe. The deepest and most spiritual leaders among us call us to love our enemies—to extend the hand of friendship to those that threaten us, to pray for those that abuse us, to love those that hate us, to treat all citizens of all countries like our neighbors.

The Spirit of God that pervades the history of the development of our species, that draws all of us slowly into the realization of the centrality of love, comes slowly to be known to us as more than simply the source of our tentatively articulated moral sentiments, or the metaphorical author of our better natures. As God is explored through our religious reflections, we understand that God is the transcendent Creator of everything and yet, remarkably, a Creator that calls us into relationship.

As Christians we celebrate God becoming incarnate in the creative order, in the fullness of time, when our species was ready. We believe that God's incarnation in a humble member of our species was motivated by his love for the world. The Creator of the universe, as revealed in Jesus, is a God of love—not political power or economic power as many would have preferred and some still do. Jesus, the most provocative and influential teacher of all time, had a singular message for his followers—to love. His followers were told to love with such clarity, in fact, that they would become known for the myriad ways in which they make this love real in the world around them.

Jesus's message summarizes and codifies our highest instincts about loving one another. To love those around us with needs—"the least of these," Jesus called them—is to love Jesus. To fail to reach out to those in need is to fail to love Jesus. "As you did it not to one of the least of these," Jesus admonished his followers, setting the bar impossibly high, "you did it not to me."[53]

Our religious and our scientific understandings of the world have at last converged on the importance of love. Perhaps we can now affirm that this is the purpose for which God called the Creation into existence and guided it into its present reality. And perhaps, in understanding this, we can go forward into an unknown and undetermined future with a vision for how we should live and what kind of future we should, in partnership with a God who calls us into fellowship, help create.

And there was evening and morning, beginning and ending, of the seventh epoch of creation. And, having entered into fellowship with creatures in the universe, God blessed the creation and rested, satisfied that the Logos, still very much present and active in the cosmos, was accomplishing its task.

And God saw that it was Good.

ACKNOWLEDGMENTS

The ideas explored in this book—*Seven Glorious Days: A Scientist Retells the Genesis Creation Story*—were originally intended to be the final chapter of my earlier book with Francis Collins, *The Language of Science and Faith: Straight Answers to Genuine Questions.* For various reasons we decided that wasn't the best idea, and the material was just set aside. Some time later I was having lunch with my friend Lil Copan—then an editor with Paraclete Press and now with Abingdon—and I mentioned to her that I had this idea about "rewriting" Genesis as a contemporary creation story. I wanted to see what that story would look like if it were freed from the "creation versus evolution" swamp in which it is often mired. Lil got quite animated and encouraged me to produce a chapter so she could get a better idea of what I was thinking. After a couple of drafts—improved with Lil's editorial input—I had something that she liked well enough to take to the marketing folk who have to worry about actually selling books. And fortunately for me, they liked it and even suggested it be expanded from the proposed length, which was basically that of a long essay.

Lil handed the editorial reins to Jon Sweeney, who edited the entire manuscript once it was completed and made many helpful suggestions. Jon also pushed me to fact-check a few claims that, it turned out, needed to be fact-checked. His input was very important.

For most of the writing of *Seven Glorious Days* I was supported by the BioLogos Foundation, founded by Francis Collins, and underwritten by grants from the Templeton Foundation and other sources. Being supported to write is an author's fantasy, of course, and I am grateful to Francis for inviting me to get involved with BioLogos. (Readers interested in the intersection of science and faith can join the discussion at www.biologos.org.)

Lastly, I would like to thank my wonderful family—my wife, Myrna, and my daughters, Sara and Laura—who have all supported and encouraged me and my irregular writing career for more years than I care to remember.

ART CREDITS

INTRODUCTION:

The Creation Story for the Twenty-First Century

http://upload.wikimedia.org/wikipedia/commons/b/b6/Gutenberg_Bible%2C_Lenox_Copy%2C_New_York_Public_Library%2C_2009._Pic_01.jpg.

DAY 1:

In the Beginning

http://upload.wikimedia.org/wikipedia/commons/7/76/Creation_of_Light_Detail_2.png.

DAY 2:

A Universe of Horseshoe Nails

http://commons.wikimedia.org/wiki/File:Principal_quantum_number_nodes.png.

DAY 3:

A Billion Stars Are Born

http://commons.wikimedia.org/wiki/File:Orion_constellation_Hevelius.jpg.

Periodic table of the elements image created by Jonathan Ramey and Karl Giberson.

DAY 4:

Looking for Life in All the Right Places

http://upload.wikimedia.org/wikipedia/commons/5/55/Bewitched_episode_1968.jpg.

http://upload.wikimedia.org/wikipedia/commons/5/5a/War-of-the-worlds-tripod.jpg.

DAY 5:

The Grand Explosion of Life

http://upload.wikimedia.org/wikipedia/commons/7/78/Earth_cutaway.png.

DAY 6:

A Child Is Born

http://www.artmagick.com/pictures/picture.aspx?id=8133&name=to-the-unknown-land.

Second image created by Karl Giberson. Based on http://filipspagnoli.files.wordpress.com/2009/07/life-expectancy-throughout-history-long-trend.gif.

THE SEVENTH, FINAL DAY OF CREATION:

The Greatest of These Is Love

http://commons.wikimedia.org/wiki/File:Rembrandt-The_return_of_the_prodigal_son.jpg.

NOTES

1 SA, JA, AB, KB, DB, RD, DS, BF, AH, MH, MJ, CJ, WJ, TL, JM, MM, JM, ZP, RP, RR, KS, SSx2, NK, AT, TV, KW, EW, MW.

2 L. Susan Stebbing, *Philosophy and the Physicists* (London: Constable, 1959), 258.

3 Arthur Eddington, *The Expanding Universe* (Cambridge: Cambridge University Press, 1933), 56 (emphasis mine).

4 Brian Swimme, *The Universe Story: From the Primordial Flaring Forth to the Ecozoic Era—A Celebration of the Unfolding of the Cosmos* (San Francisco: HarperOne, 1994).

5 Kathy Sawyer, "'Big Bang' Remains That as Astronomers' Renaming Contest Implodes," *Washington Post*, January 14, 1994.

6 Robert Jastrow, *God and the Astronomers* (New York: W. W. Norton & Company, 1978), 116.

7 Freeman Dyson, *Disturbing the Universe* (New York: Harper & Row, 1979), 250–51.

8 Arthur Eddington, *Space, Time, and Gravitation: An Outline of the General Relativity Theory* (Cambridge: Cambridge University Press, 1921), 201.

9 Pierre-Simon Laplace, *A Philosophical Essay on Probabilities*, trans. F. W. Truscott and F. L. Emory (New York: Dover Publications, 1951), 4.

10 As I write this, some recent experiments at the CERN laboratories in Europe have suggested that a certain type of neutrino has been observed going a bit faster than light. The resolution of this controversial possibility will be interesting to watch. See Eryn Brown and Amina Khan, "Faster than Light? CERN Findings Bewilder Scientists," *Los Angeles Times*, September 23, 2011.

11 Robert P. Crease and Charle C. Mann, *The Second Creation: Makers of the Revolution in Twentieth-Century Physics* (New York: Macmillan, 1986), 282.

12 This mathematical analogy is one that I created for a blog I wrote: "Mathematics and the Religious Impulse," *Huffington Post*, August 8, 2010, http://www.huffingtonpost.com/karl-giberson-phd/mathematics-and-the-relig_b_673359.html. I also used the analogy in my earlier book *The Wonder of the Universe: Hints of God in Our Fine-Tuned World* (Downers Grove, IL: InterVarsity Press, 2012). I trust the reader will forgive me for plagiarizing myself.

13 David Park, *The Grand Contraption: The World as Myth, Number, and Chance* (Princeton: Princeton University Press, 2007), 37.

14 Ibid.

15 Timothy Ferris, *Coming of Age in the Milky Way,* (New York: William Morrow, 1988), 61.

16 John Robert Christianson, *On Tycho's Island: Tycho Brahe, Science, and Culture in the Sixteenth Century* (New York: Cambridge University Press, 2003), 18.

17 I am ignoring the small electromagnetic interaction that creates hydrogen molecules out of hydrogen atoms. This is not essential for this part of our story.

18 Portions of this discussion are adapted from a blog I wrote: "The Beauty of Being a Scientist and a Christian," *Huffington Post,* April 21, 2010, http://www.huffingtonpost.com/karl-giberson-phd/the-beauty-of-being-a-sci_b_546062.html.

19 Carl Sagan, *Pale Blue Dot: A Vision of the Human Future in Space.* (New York: Random House, 1994), 9.

20 David Park, *The Grand Contraption*, 209.

21 http://upload.wikimedia.org/wikipedia/commons/5/55/Bewitched_episode_1968.jpg.
http://upload.wikimedia.org/wikipedia/commons/5/5a/War-of-the-worlds-tripod.jpg.

22 "Charles Darwin & Evolution 1809–2009," Christ's College, Cambridge, http://www.christs.cam.ac.uk/darwin200/pages/index.php?page_id=f8.

23 Bill Bryson, *A Short History of Nearly Everything: Special Illustrated Edition* (New York: Broadway Books, 2010), 290.

24 Ibid., 291.

25 Fred Hoyle, *The Intelligent Universe* (London: Michael Joseph, 1983), 19.

26 Gould quoted in Bill Bryson, *A Short History of Nearly Everything* (New York: Broadway Books, 2003), 367.

27 Paul Davies, *The Fifth Miracle: The Search for the Origin and Meaning of Life* (New York: Touchstone, 1999), 22.

28 Ibid., 92.

29 Ibid., 19.

30 Edward O. Wilson, *The Diversity of Life* (Cambridge, MA: Harvard University Press, 1992), 35.

31 From the website of Answers in Genesis, a group claiming to "relate the relevance of a literal Genesis to the church and the world today with creativity." See "'People For the American Way' Betray Their Name!" April 17, 2000, http://www.answersingenesis.org/articles/2000/04/17/betray-their-name.

32 Robert Wright, *NonZero: The Logic of Human Destiny* (New York: Vintage Books, 2000), 254.

33 Ibid., 252.

34 Deborah Blum, *Love at Goon Park: Harry Harlow and the Science of Affection* (Cambridge, MA: Perseus Publishing, 2004), 31–32.

35 William Farr, *First Annual Report of the Registrar-General of Births, Deaths, and Marriages, in England,* 19th Century House of Commons Sessional Papers 1839: 44, quoted in Jon Greene, "What Was Life Like for Children in Victorian England?" *Helium,* October 24, 2011, http://www.helium.com/items/2245171-victorian-childhood.

36 See the U.S. Census Bureau's "World POPClock Project," http://www.census.gov/population/popclockworld.html. The figure cited was projected by the clock on May 16, 2012.

37 "Fun Facts about the Brain," *Brain Health and Puzzles,* http://www.brainhealthandpuzzles.com/fun_facts_about_the_brain.html.

38 Wigner, E. P. "*The unreasonable effectiveness of mathematics in the natural sciences. Richard Courant lecture in mathematical sciences delivered at New York University, May 11, 1959.*" *Communications on Pure and Applied Mathematics* 13: 1–14. doi:10.1002/cpa.3160130102.

39 Penrose, Roger, *Shadows of the Mind: A Search for the Missing Science of Consciousness* (New York: Oxford University Press, 1994) p. 50-51.

40 http://rescomp.standford.edu/~cheshire/EinsteinQuotes.html (accessed May 15, 2012).

41 http://www.gutenberg.org/catalog/world/readfile?fk-files=89&pageno=39.

42 1 John 4:7–8.

43 Daniel Ladinsky, trans., "The Stairway of Existence," in *The Gift: Poems by the Great Sufi Master* (New York: Penguin Books, 1999), 96.

44 http://www-history.mcs.st-and.ac.uk/Quotations/Einstein.html (accessed March 30, 2012).

45 This account of Harry Harlow's revolutionary ideas about the importance of loving infants is based loosely on my essay "The Greatest of These," in the July/August 2005 issue of *Books & Culture*. The essay is available on my website: http://www.karlgiberson.com/writing. The full story can be found in Deborah Blum's superb biography *Love at Goon Park*.

46 Steven Weinberg, *The First Three Minutes: A Modern View of the Origin of the Universe*, updated ed. (New York: Basic Books, 1993), 154.

47 Steven Weinberg, *Dreams of a Final Theory: The Scientist's Search for the Ultimate Laws of Nature* (New York: Pantheon Books, 1992), 245.

48 Dyson, *Disturbing the Universe*, 250.

49 Freeman Dyson, *Infinite in All Directions* (New York: Harper & Row, 1988) 297.

50 Freeman Dyson, "Progress in Religion" (remarks delivered at event celebrating his receipt of the 2000 Templeton Prize). http://www.edge.org/documents/archive/edge68.html (accessed March 30, 2012).

51 John Polkinghorne, *Theology in the Context of Science* (New Haven: Yale University Press, 2010), 113.

52 Ibid., 114.

53 Matthew 25:45.

ABOUT PARACLETE PRESS

Who We Are

Paraclete Press is a publisher of books, recordings, and DVDs on Christian spirituality. Our publishing represents a full expression of Christian belief and practice—from Catholic to Evangelical, from Protestant to Orthodox.

We are the publishing arm of the Community of Jesus, an ecumenical monastic community in the Benedictine tradition. As such, we are uniquely positioned in the marketplace without connection to a large corporation and with informal relationships to many branches and denominations of faith.

What We Are Doing

Books Paraclete publishes books that show the richness and depth of what it means to be Christian. Although Benedictine spirituality is at the heart of all that we do, we publish books that reflect the Christian experience across many cultures, time periods, and houses of worship. We publish books that nourish the vibrant life of the church and its people—books about spiritual practice, formation, history, ideas, and customs.

We have several different series, including the best-selling Paraclete Essentials and Paraclete Giants series of classic texts in contemporary English; A Voice from the Monastery—men and women monastics writing about living a spiritual life today; award-winning poetry; best-selling gift books for children on the occasions of baptism and first communion; and the Active Prayer Series that brings creativity and liveliness to any life of prayer.

Recordings From Gregorian chant to contemporary American choral works, our music recordings celebrate sacred choral music through the centuries. Paraclete distributes the recordings of the internationally acclaimed choir Gloriæ Dei Cantores, praised for their "rapt and fathomless spiritual intensity" by *American Record Guide*, and the Gloriæ Dei Cantores Schola, which specializes in the study and performance of Gregorian chant. Paraclete is also the exclusive North American distributor of the recordings of the Monastic Choir of St. Peter's Abbey in Solesmes, France, long considered to be a leading authority on Gregorian chant.

Videos Our videos offer spiritual help, healing, and biblical guidance for life issues: grief and loss, marriage, forgiveness, anger management, facing death, and spiritual formation.

Learn more about us at our website:
www.paracletepress.com, or call us toll-free at 1-800-451-5006.